Electrochemistry

Nanoparticles and Nanomaterials: Designs, Characterization and Applications

Editor: Andrew Green

NY RESEARCH PRESS

New York

Published by NY Research Press
118-35 Queens Blvd., Suite 400,
Forest Hills, NY 11375, USA
www.nyresearchpress.com

Nanoparticles and Nanomaterials: Designs, Characterization and Applications
Edited by Andrew Green

International Standard Book Number: 978-1-63238-645-8 (Hardback)

Cataloging-in-Publication Data

Nanoparticles and nanomaterials : designs, characterization and applications / edited by Andrew Green.
p. cm.
Includes bibliographical references and index.
ISBN 978-1-63238-645-8
1. Nanoparticles. 2. Nanostructured materials. 3. Nanostructured materials--
Design and construction. I. Green, Andrew.
TA418.9.N35 N36 2019
620.5--dc23

Contents

Preface...VII

Chapter 1   **Anti-inflammatory effect of fullerene C$_{60}$ in a mice model of atopic dermatitis**............ 1
Nadezda Shershakova, Elena Baraboshkina, Sergey Andreev, Daria Purgina,
Irina Struchkova, Oleg Kamyshnikov, Alexandra Nikonova and Musa Khaitov

Chapter 2   **A versatile papaya mosaic virus (PapMV) vaccine platform based on
sortase-mediated antigen coupling**...12
Ariane Thérien, Mikaël Bédard, Damien Carignan, Gervais Rioux,
Louis Gauthier-Landry, Marie-Ève Laliberté-Gagné, Marilène Bolduc,
Pierre Savard and Denis Leclerc

Chapter 3   **Real-time, label-free monitoring of cell viability based on cell adhesion
measurements with an atomic force microscope**...25
Fang Yang, René Riedel, Pablo del Pino, Beatriz Pelaz, Alaa Hassan Said,
Mahmoud Soliman, Shashank R. Pinnapireddy, Neus Feliu,
Wolfgang J. Parak, Udo Bakowsky and Norbert Hampp

Chapter 4   **Comparative efficacy analysis of anti-microbial peptides, LL-37 and
indolicidin upon conjugation with CNT, in human monocytes**.................................35
Biswaranjan Pradhan, Dipanjan Guha, Krushna Chandra Murmu, Abhinav Sur,
Pratikshya Ray, Debashmita Das and Palok Aich

Chapter 5   **Enzyme adsorption-induced activity changes: a quantitative study on TiO$_2$
model agglomerates**..51
Augusto Márquez, Krisztina Kocsis, Gregor Zickler, Gilles R. Bourret,
Andrea Feinle, Nicola Hüsing, Martin Himly, Albert Duschl, Thomas Berger and
Oliver Diwald

Chapter 6   **A novel covalent approach to bio-conjugate silver coated single walled carbon
nanotubes with antimicrobial peptide**..61
Atul A. Chaudhari, D'andrea Ashmore, Subrata deb Nath, Kunal Kate,
Vida Dennis, Shree R. Singh, Don R. Owen, Chris Palazzo, Robert D. Arnold,
Michael E. Miller and Shreekumar R. Pillai

Chapter 7   **Evaluation of the antibacterial power and biocompatibility of zinc oxide
nanorods decorated graphene nanoplatelets: new perspectives for
antibiodeteriorative approaches**..76
Elena Zanni, Erika Bruni, Chandrakanth Reddy Chandraiahgari,
Giovanni De Bellis, Maria Grazia Santangelo, Maurizio Leone,
Agnese Bregnocchi, Patrizia Mancini, Maria Sabrina Sarto and Daniela Uccelletti

Chapter 8   **The role of intracellular trafficking of CdSe/ZnS QDs on their consequent
toxicity profile**..88
Bella B. Manshian, Thomas F. Martens, Karsten Kantner, Kevin Braeckmans,
Stefaan C. De Smedt, Jo Demeester, Gareth J. S. Jenkins, Wolfgang J. Parak,
Beatriz Pelaz, Shareen H. Doak, Uwe Himmelreich and Stefaan J. Soenen

Chapter 9   **Characteristics and properties of nano-LiCoO$_2$ synthesized by pre-organized single source precursors: Li-ion diffusivity, electrochemistry and biological assessment**..**102**
Jean-Pierre Brog, Aurélien Crochet, Joël Seydoux, Martin J. D. Clift, Benoît Baichette, Sivarajakumar Maharajan, Hana Barosova, Pierre Brodard, Mariana Spodaryk, Andreas Züttel, Barbara Rothen-Rutishauser, Nam Hee Kwon and Katharina M. Fromm

Chapter 10   **Toxicity of nano- and ionic silver to embryonic stem cells: a comparative toxicogenomic study**..**125**
Xiugong Gao, Vanessa D. Topping, Zachary Keltner, Robert L. Sprando and Jeffrey J. Yourick

Chapter 11   **Synthesis and characterization of crosslinked polyisothiouronium methylstyrene nanoparticles of narrow size distribution for antibacterial and antibiofilm applications**...**143**
Sarit Cohen, Chen Gelber, Michal Natan, Ehud Banin, Enav Corem-Salkmon and Shlomo Margel

Chapter 12   **Nanostructured biosensor using bioluminescence quenching technique for glucose detection**...**153**
Longyan Chen, Longyi Chen, Michelle Dotzert, C. W. James Melling and Jin Zhang

Chapter 13   **A *retro-inverso* cell-penetrating peptide for siRNA delivery** **162**
Anaïs Vaissière, Gudrun Aldrian, Karidia Konate, Mattias F. Lindberg, Carole Jourdan, Anthony Telmar, Quentin Seisel, Frédéric Fernandez, Véronique Viguier, Coralie Genevois, Franck Couillaud, Prisca Boisguerin and Sébastien Deshayes

Chapter 14   **Photoinduced effects of m-tetrahydroxyphenylchlorin loaded lipid nanoemulsions on multicellular tumor spheroids**..................................**180**
Doris Hinger, Fabrice Navarro, Andres Käch, Jean-Sébastien Thomann, Frédérique Mittler, Anne-Claude Couffin and Caroline Maake

Chapter 15   **Cell-based cytotoxicity assays for engineered nanomaterials safety screening: exposure of adipose derived stromal cells to titanium dioxide nanoparticles**..**194**
Yan Xu, M. Hadjiargyrou, Miriam Rafailovich and Tatsiana Mironava

Chapter 16   **Understanding cellular internalization pathways of silicon nanowires**...................................**211**
Kelly McNear, Yimin Huang and Chen Yang

Chapter 17   **Alternative moth-eye nanostructures: antireflective properties and composition of dimpled corneal nanocoatings in silk-moth ancestors**...................**221**
Mikhail Kryuchkov, Jannis Lehmann, Jakob Schaab, Vsevolod Cherepanov, Artem Blagodatski, Manfred Fiebig and Vladimir L. Katanaev

Permissions

List of Contributors

Index

Preface

In my initial years as a student, I used to run to the library at every possible instance to grab a book and learn something new. Books were my primary source of knowledge and I would not have come such a long way without all that I learnt from them. Thus, when I was approached to edit this book; I became understandably nostalgic. It was an absolute honor to be considered worthy of guiding the current generation as well as those to come. I put all my knowledge and hard work into making this book most beneficial for its readers.

Nanoparticles and nanomaterials are studied in the interdisciplinary science of nanotechnology. Nanomaterials are particles ranging in size between 1 and 100 nm. Nanostructures are structures intermediate between microscopic and molecular structures. The study of these is achieved by integrating the principles of semiconductor physics, surface science, organic chemistry, molecular engineering, etc. The applications of nanoparticles and nanostructures are diverse. Prominent among these are nanomedicine, nanoelectronics, biomaterials, biosensors, etc. Research in this domain mostly explores new varieties of nanoparticle design and its applications. This book has been compiled to provide a detailed overview of the design and characterization of nanoparticles with emphasis on their applications. It brings forth some of the most innovative concepts and elucidates the unexplored aspects of nanotechnology. It aims to serve as a resource guide for engineers, physicists, material scientists and students.

I wish to thank my publisher for supporting me at every step. I would also like to thank all the authors who have contributed their researches in this book. I hope this book will be a valuable contribution to the progress of the field.

Editor

Electrochemistry

Margot Reilly

WILLFORD PRESS
www.willfordpress.com

Published by Willford Press,
118-35 Queens Blvd., Suite 400,
Forest Hills, NY 11375, USA

ISBN: 978-1-68285-997-1

Cataloging-in-Publication Data

Electrochemistry / Margot Reilly.
    p. cm.
Includes bibliographical references and index.
ISBN 978-1-68285-997-1
1. Electrochemistry. 2. Chemistry, Physical and theoretical. I. Reilly, Margot.
QD553 .E44 2022
541.37--dc23

For information on all Willford Press publications
visit our website at www.willfordpress.com

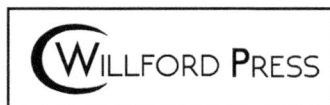

WILLFORD PRESS

TABLE OF CONTENTS

Preface VII

Chapter 1   Introduction 1

- Electrochemical Cell   6
- Voltaic Cell   8
- Nernst Equation   12
- Equilibrium Constant   14
- Mass Transfer   18
- Cottrell Equation   23
- Electrode Potential   23

Chapter 2   Electrochemical Reactions 27

- Electrochemical Reaction Mechanism   30
- Redox Reaction   32
- Reduction Potential   41
- Half-reaction   45
- Combination Reaction   48
- Corrosion Reaction   51
- Applications of Electrochemical Reaction   94

Chapter 3   Electrochemical Engineering 98

- Electrochemical Reduction of Carbon Dioxide   98
- Fuel Cell   100
- Electrochemical Gas Sensor   118
- Flow Battery   119
- Electrochemical Machining   126
- Electrophoresis   130

Chapter 4   Electrolysis 133

- Electrolytic Cell   135
- Hofmann Voltameter   139
- Faradays Laws of Electrolysis   140

- Principle of Electrolysis of Copper Sulfate Electrolyte   142
- Anodizing   146
- Castner Process   153
- Castner–Kellner Process   154
- Chloralkali Process   156
- Hall–Héroult Process   159
- Electrolysis of Water   163
- Polymer Electrolyte Membrane Electrolysis   172
- Electrowinning   177
- High-temperature Electrolysis   179
- Kolbe Electrolysis   181
- Pulse Electrolysis   182
- Patterson Power Cell   187
- Applications of Electrolysis   189

Chapter 5   **Electroanalytical Methods** **194**

- Coulometry   195
- Voltammetry   199
- Amperometric Titration   217
- Electrochemical Stripping Analysis   218
- Chronoamperometry   219
- Potentiometry   223
- Electrogravimetry   224

Permissions

Index

The purpose of this book is to help students understand the fundamental concepts of this discipline. It is designed to motivate students to learn and prosper. I am grateful for the support of my colleagues. I would also like to acknowledge the encouragement of my family.

The branch of physical chemistry which deals with the study of the relationship between electricity as a quantitative and measurable phenomena, and chemical change, is referred to as electrochemistry. One of the major areas of study within this field is redox reactions. These reactions involve the transfer of electrons between molecules or atoms. Electrochemistry is used to describe overall reactions in which the separate redox reactions are connected by an external electric circuit and an intervening electrolyte. It also deals with balancing redox reactions. This book outlines the concepts and applications of electrochemistry in detail. While understanding the long-term perspectives of the topics, it makes an effort in highlighting their impact as a modern tool for the growth of the discipline. This book is appropriate for students seeking detailed information in this area as well as for experts.

A foreword for all the chapters is provided below:

Chapter – Introduction

The branch of physical chemistry that studies the relationship between electricity and identifiable chemical change is known as electrochemistry. The main areas of study in electrochemistry are photochemistry and quantum electrochemistry. This is an introductory chapter which will introduce briefly all the significant types and aspects of electrochemistry.

Chapter – Electrochemical Reactions

The processes in which electrons flow between solid electrodes and a substance, known as electrolyte, are known as electrochemical reactions. Some of the examples of reactions are redox reactions, corrosion reaction and combination reaction. The chapter closely examines these key electrochemical reactions to provide an extensive understanding of the subject.

Chapter – Electrochemical Engineering

The domain within chemical engineering which focuses on the technological applications of electrochemical phenomena is referred to as electrochemical engineering. Electrochemical gas sensor, glow battery, electrophoresis, etc. are some of the concepts that fall in its domain. This chapter discusses in detail these concepts related to electrochemical engineering.

Chapter – Electrolysis

to drive a non-spontaneous chemical reaction by using a direct electric current is known as electrolysis. Some of the processes that it includes are Castner–Kellner process, Castner process, chloralkali process and Hall-Heroult process. The chapter closely examines these key concepts and processes of electrolysis to provide an extensive understanding of the subject.

Chapter – Electroanalytical Methods

The techniques in analytical chemistry that deal with the study of an analyte is known as electroanalytical method. It measures the potential and current present in an electrochemical cell. The main electroanalytical method are coulometry and voltammetry. The topics elaborated in this chapter will help in gaining a better perspective about electroanalytical method.

Margot Reilly

CHAPTER 1

Introduction

The branch of physical chemistry that studies the relationship between electricity and identifiable chemical change is known as electrochemistry. The main areas of study in electrochemistry are photochemistry and quantum electrochemistry. This is an introductory chapter which will introduce briefly all the significant types and aspects of electrochemistry.

Electrochemistry deals with the links between chemical reactions and electricity. This includes the study of chemical changes caused by the passage of an electric current across a medium, as well as the production of electric energy by chemical reactions. Electrochemistry also embraces the study of electrolyte solutions and the chemical equilibria that occur in them.

Many chemical reactions require the input of energy. Such reactions can be carried out at the surfaces of electrodes in cells connected to external power supplies. These reactions provide information about the nature and properties of the chemical species contained in the cells, and can also be used to synthesize new chemicals. The production of chlorine and aluminum and the electroplating and electrowinning of metals are examples of industrial electrochemical processes. Electrochemical cells that produce electric energy from chemical energy are the basis of primary and secondary (storage) batteries and fuel cells. Other electrical phenomena of interest in chemical systems include the behavior of ionic solutions and the conduction of current through these solutions, the separation of ions by an electric field (electrophoresis), the corrosion and passivation of metals, electrical effects in biological systems (bioelectrochemistry), and the effect of light on electrochemical cells (photoelectrochemistry).

Electrochemical Cells

An electrochemical cell generally consists of two half-cells, each containing an electrode in contact with an electrolyte. The electrode is an electronic conductor (such as a metal or carbon) or a semiconductor. Current flows through the electrodes via the movement of electrons. An electrolyte is a phase in which charge is carried by ions. For example, a solution of table salt (sodium chloride, $NaCl$) in water is an electrolyte containing sodium cations (Na^+) and chloride anions (Cl^-). When an electric field is applied across this solution, the ions move: Na^+ toward the negative side of the field and Cl^- toward the positive side.

The half-cells are connected by a cell separator that allows ions to move between the half-cells but prevents mixing of the electrolytes. The separator can consist of a salt bridge, or tube of aqueous solution plugged at both ends with glass wool, or it can be an ion exchange membrane or a sintered-glass disk. In some cases both half-cells use the same electrolyte, so that the electrochemical cell consists of two electrodes in contact with a single electrolyte. Electrochemical cells are usually classified as either galvanic or electrolytic. In galvanic cells, reactions occur spontaneously at the electrode–electrolyte interfaces when the two electrodes are connected by a conductor such as a metal wire. Galvanic cells convert chemical energy to electric energy and are the components of batteries, which usually contain several cells connected in series. In electrolytic cells, reactions are forced to occur at the electrode–electrolyte interfaces by way of an external source of power connected to both electrodes. Electric energy from the external source is converted to chemical energy in the form of the products of the electrode reactions.

The galvanic cell shown in the figure is known as the Daniell cell and was used as an early source of energy. It consists of a zinc (Zn) electrode in contact with an aqueous zinc sulfate solution and a copper (Cu) electrode in contact with an aqueous copper sulfate solution. When the external switch is closed, an atom of zinc on the zinc electrode is oxidized to zinc ion, liberating two electrons.

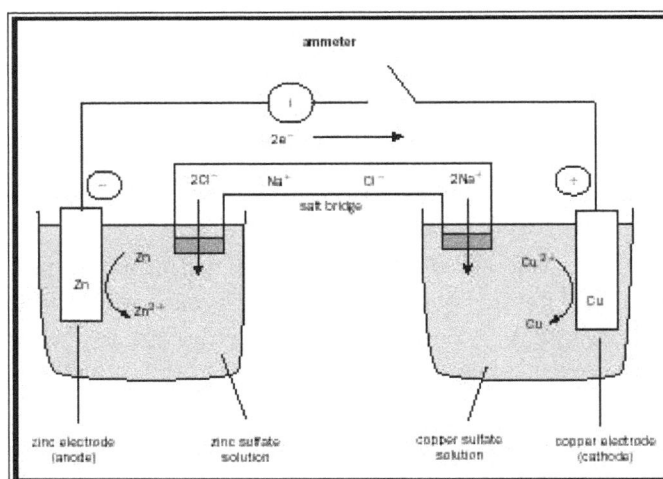

$$Zn \rightarrow Zn^{2+} + 2e^-$$

Daniell Cell.

The electrons pass through the external wire and reduce a copper ion to an atom of copper metal on the surface of the copper electrode.

$$Cu^{2+} + 2e^- \rightarrow Cu$$

The electron flow in the external circuit represents an electric current produced by the cell. Ions flow within the electrolytes and across the salt bridge, as shown in the figure, to prevent an imbalance of ionic charge in the solutions that could result from the occurrence of these two electrode reactions. The overall cell reaction is the reduction of copper ion by zinc.

$$Cu^{2+} + Zn \rightarrow Cu + Zn^{2+}$$

The electrolytic cell shown in the figure is the industrial chloralkali cell in which brine (an aqueous sodium chloride solution) is electrolytically converted to chlorine and caustic soda (sodium hydroxide, NaOH). The external power source supplies electric energy to drive the overall reaction.

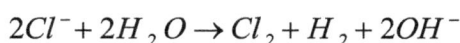

$$2Cl^- + 2H_2O \rightarrow Cl_2 + H_2 + 2OH^-$$

Chloride ion is oxidized to chlorine gas at the carbon electrode, and water is reduced to hydrogen gas (H_2) and hydroxide ion (OH^-) at the iron electrode. The electrolytes are maintained as electrically neutral by a flow of sodium ions through the separator (such as an ion exchange membrane).

The electrode where oxidation occurs, the zinc electrode in figure and the carbon electrode in the figure, is called the anode, while the electrode where reduction occurs is called the cathode. Reactions $Zn \rightarrow Zn^{2+} + 2e^-$ and $Cu^{2+} + 2e^- \rightarrow Cu$ are known as half-reactions, whereas reactions $Cu^{2+} + Zn \rightarrow Cu + Zn^{2+}$ and $2Cl^- + 2H_2O \rightarrow Cl_2 + H_2 + 2OH^-$ are called oxidation-reduction (redox) reactions.

Electrode Potentials

Current and potential (or voltage) are the two electrical variables of greatest interest in electrochemical cells. Current is related to the rate of the electrode reactions, and the potential, to the cell energetics. Current is measured in amperes (A), or the amount of electricity in coulombs (C) that passes across a medium per second. Potential between the two electrodes is measured in volts (V) with a voltmeter. Potential (V) has units of energy or work (joules, J) per amount of electric charge (C). That is, 1 V = 1 J/C, so that the cell potential is a measure of the energy of the cell reaction. The cell is said to be at open circuit when no current flows; that is, when there are no external connections to the electrodes. Under these conditions, no electrode reactions occur.

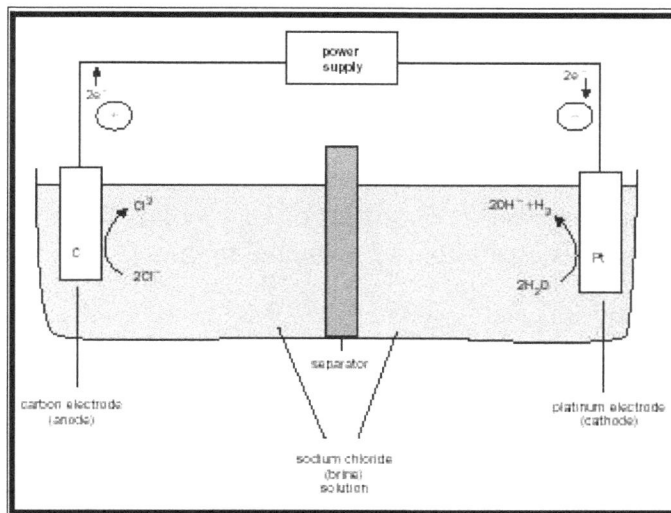

Electrolytic Cell.

Measurements of the potentials of galvanic cells at open circuit give information about the thermodynamics of cells and cell reactions. For example, the potential of the cell in figure, when the solution concentrations are 1 molar (1 M) at 25 °C, is 1.10 V. This is called the standard potential of the cell and is represented by E°. The available energy (the Gibb's free energy $\Delta G°$) of the cell reaction given in equation $Cu^{2+} + Zn \rightarrow Cu + Zn^{2+}$ is related to E° by:

$$\Delta G° = -nF E°$$

Where n is the number of electrons transferred in the reaction (in this case two) and F is a proportionality constant, called the Faraday (96,485 coulombs/equivalent). The cell potential is the difference in potential of the two half-cells. Tables of standard electrode potentials of half-reactions have been compiled; representative values are given in table. These are frequently tabulated with respect to the standard or normal hydrogen electrode (SHE or NHE), which is arbitrarily assigned a half-cell potential of zero. Thus the value, +0.34 V, is assigned to the half-reaction $Cu^{2+} + 2e^- \rightarrow Cu$ and,

$$E° = +0.34 \text{ V vs NHE}$$

Similarly, the standard potential for the Zn/Zn^{2+} cell yields $Zn^{2+} + 2e^- \rightarrow Zn$ and

Table: Representative Standard Potentials.

Half Reaction	E °vs NHE
$Li^+ + e^- \rightarrow Li$	−3.045
$Mg^{2+} + 2e^- \rightarrow Mg$	−2.356
$Al^{3+} + 3e^- \rightarrow Al$	−1.67
$Zn^{2+} + 2e^- \rightarrow Zn$	−0.7626
$Cr^{3+} + e^- \rightarrow Cr^{2+}$	−0.424
$2H^+ + 2e^- \rightarrow H_2$	0.000
$Cu^{2+} + 2e^- \rightarrow Cu$	0.340
$O_2 + 4H^+ + 4e^- \rightarrow 2H_2O$	1.229
$F_2 + 2e^- \rightarrow 2F^-$	2.87

$$E° = -0.76 \text{ V vs NHE}$$

The difference between these two half-cell potentials yields the standard potential of the Zn-Cu cell.

The standard potential applies to a half-cell when all the reactants are present at unit activity; that is, when the solution species are near a concentration of 1 molar. The actual half-cell potential E is a function of the solution concentrations and is related to these and to the standard potential E° by the Nernst equation.

Common Batteries

In most flashlights, toys, and remote controllers for televisions, primary batteries are used. The cell reactions in primary batteries are irreversible. During use, reactants are converted to products, and when the reactants are used up, the battery is "dead". The inexpensive flashlight batteries sold in retail stores use a design called a Leclanche dry cell. The body of the battery is made of zinc, which acts as the anode. A carbon rod in the center of the cell serves as the cathode. It is surrounded by a moist paste of graphite powder (carbon), manganese dioxide (MnO_2) and ammonium chloride (NH_4Cl). The anode reaction is the oxidation of the zinc cylinder to zinc ions. The cathode reaction involves the reduction of manganese dioxide. A simplified version of the overall reaction is:

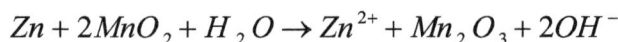

$$Zn + 2MnO_2 + H_2O \rightarrow Zn^{2+} + Mn_2O_3 + 2OH^-$$

Alkaline cells are similar, except that the zinc case is porous and the paste around the carbon cathode is moist manganese dioxide and potassium hydroxide. These are more expensive than ordinary zinc-carbon cells, but they maintain a high voltage longer.

The lead-acid storage battery used in automobiles is a secondary battery; it is rechargeable. That is, the automobile battery operates as a galvanic cell when used to start the engine (when discharging), and as an electrolytic cell when it is charged by the alternator or by an external battery charger. The anode consists of porous lead plates in contact with a sulfuric acid (H_2SO_4) solution. The cathode consists of lead dioxide (PbO_2) plates, also in sulfuric acid. Electrons flow from the lead plates to the lead oxide plates. As lead (Pb) loses electrons, it forms lead ions (Pb^{2+}) that react with sulfate ions (SO_4^{2-}) in solution to form insoluble lead sulfate ($PbSO_4$). When PbO_2 gains electrons, it too reacts with SO_4^{2-} ions in solution to form solid $PbSO_4$. The cell reaction is:

$$Pb + PbO_2 + 4H^+ + 2SO_4^{2-} \rightarrow 2PbSO_4 + 2H_2O$$

and proceeds from left to right when the battery is discharging and from right to left when charging.

The rechargeable nickel-cadmium (Ni-Cad) batteries are used in a variety of cordless appliances such as telephones, battery operated tools, and portable computers. During discharge, cadmium metal (Cd) acts as the anode, and nickel dioxide (NiO_2) as the cathode. Both metals form insoluble hydroxides due to the presence of the potassium hydroxide electrolyte. The cell reaction during discharge is:

$$F_i = -\underbrace{RT\nabla \ln a_i}_{\text{Chemical}} - \underbrace{z_iF\nabla \phi}_{\text{Electrical}}$$

The reaction is reversed during charging.

Photoelectrochemistry

Photoelectrochemistry refers to a scientific field of study that evaluates the relationship between light and electrochemical systems, including the corrosion that occurs on a metallic surface that is exposed to sunlight.

The field of photoelectrochemistry plays a role in safeguarding metallic structures that are installed outdoors with periodic or regular exposure to ultraviolet (UV) sunlight emissions.

Empirical investigations have shown that a relationship exists between light and electrochemical cells, which in turn affects the corrosion rate. In particular, ultraviolet light accelerates the corrosion of metallic surfaces.

Metals that corrode faster due to sunlight can be stabilized to reduce corrosion by using a sacrificial anode or secondary metal.

Quantum Electrochemistry

Quantum electrochemistry is use of quantum mechanical tools including density functional theory towards the study of electrochemical functions, including electron move at electrodes. It also includes models including Marcus theory. Most of the time, the field contains the notions arising in electrodynamics, quantum technicians, and electrochemistry; so is studied by way of very large variety of different professional researchers.

ELECTROCHEMICAL CELL

Electrochemical cells are nothing but devices that are capable of converting chemical energy into electrical energy, or vice versa. A common example of an electrochemical cell is a standard 1.5-volt cell which is used to power many electrical appliances such as TV remotes and clocks.

Such cells capable of generating an electric current from the chemical reactions occurring in them care called Galvanic cells or Voltaic cells. Alternatively, the cells which cause chemical reactions to occur in them when an electric current is passed through them are called electrolytic cells.

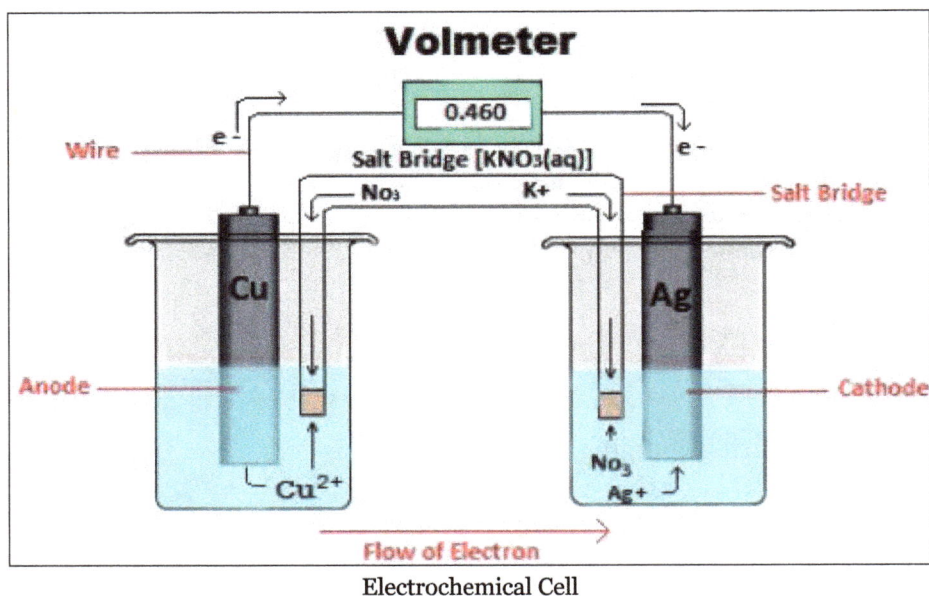

Electrochemical Cell

Electrochemical cells generally consist of a cathode and an anode. The key features of the cathode and the anode are listed in the tabular column provided in the table below.

Cathode	Anode
Denoted by a positive sign since electrons are consumed here	Denoted by a negative sign since electrons are liberated here
A reduction reaction occurs in the cathode of an electrochemical cell	An oxidation reaction occurs here
Electrons move into the cathode	Electrons move out of the anode

General convention dictates that the cathode must be represented on the right-hand side whereas the anode is represented on the left-hand side while denoting an electrochemical cell.

Half-cells and Cell Potential

- Electrochemical Cells are made up of two half-cells, each consisting of an electrode which is dipped in an electrolyte. The same electrolyte can be used for both half cells.

- These half cells are connected by a salt bridge which provides the platform for ionic contact between them without allowing them to mix with each other. An example of a salt bridge is a filter paper which is dipped in a potassium nitrate or sodium chloride solution.

- One of the half cells of the electrochemical cell loses electrons due to oxidation and the other gains electrons in a reduction process. It can be noted that an equilibrium reaction occurs in both the half cells, and once the equilibrium is reached, the net voltage becomes 0 and the cell stops producing electricity.

- The tendency of an electrode which is in contact with an electrolyte to lose or gain electrons is described by its electrode potential. The values of these potentials can be used to predict the overall cell potential. Generally, the electrode potentials are measured with the help of the standard hydrogen electrode as a reference electrode (an electrode of known potential).

Primary and Secondary Cells

- Primary cells are basically use-and-throw galvanic cells. The electrochemical reactions that take place in these cells are irreversible in nature. Hence, the reactants are consumed for the generation of electrical energy and the cell stops producing an electric current once the reactants are completely depleted.

- Secondary cells (also known as rechargeable batteries) are electrochemical cells in which the cell has a reversible reaction, i.e. the cell can function as a Galvanic cell as well as an Electrolytic cell.

- Most of the primary batteries (multiple cells connected in series, parallel, or a combination of the two) are considered wasteful and environmentally harmful devices. This is because they require about 50 times the energy they contain in their manufacturing process. They also contain many toxic metals and are considered to be hazardous waste.

Types of Electrochemical Cell

There are two types of electrochemical cells:

- Galvanic cells (also known as Voltaic cells).

- Electrolytic cell.

The key differences between Galvanic cells and electrolytic cells are tabulated below:

Galvanic Cell / Voltaic Cell	Electrolytic Cell
Chemical energy is transformed into electrical energy in these electrochemical cells.	Electrical energy is transformed into chemical energy in these cells.
The redox reactions that take place in these cells are spontaneous in nature.	An input of energy is required for the redox reactions to proceed in these cells, i.e. the reactions are non-spontaneous.
In these electrochemical cells, the anode is negatively charged and the cathode is positively charged.	These cells feature a positively charged anode and a negatively charged cathode.
The electrons originate from the species that undergoes oxidation.	Electrons originate from an external source (such as a battery).

Applications of Electrochemical Cell

- Electrolytic cells are used in the electrorefining of many non-ferrous metals. They are also used in the electrowinning of these metals.

- The production of high-purity lead, zinc, aluminium, and copper involves the use of electrolytic cells.

- Metallic sodium can be extracted from molten sodium chloride by placing it in an electrolytic cell and passing an electric current through it.

- Many commercially important batteries (such as the lead-acid battery) are made up of Galvanic cells.

- Fuel cells are an important class of electrochemical cells that serve as a source of clean energy in several remote locations.

VOLTAIC CELL

A voltaic cell is a device that produces an electric current from energy released by a spontaneous redox reaction in two half-cells.

An electrochemical cell is a device that produces an electric current from energy released by a spontaneous redox reaction. This kind of cell includes the galvanic, or voltaic, cell, named after Luigi Galvani and Alessandro Volta. These scientists conducted several experiments on chemical reactions and electric current during the late 18th century.

Electrochemical cells have two conductive electrodes, called the anode and the cathode. The anode is defined as the electrode where oxidation occurs. The cathode is the electrode where reduction takes place. Electrodes can be made from any sufficiently conductive materials, such as metals, semiconductors, graphite, and even conductive polymers. In between these electrodes is the electrolyte, which contains ions that can freely move.

The voltaic cell uses two different metal electrodes, each in an electrolyte solution. The anode will undergo oxidation and the cathode will undergo reduction. The metal of the anode will oxidize, going from an oxidation state of 0 (in the solid form) to a positive oxidation state, and it will become an ion. At the cathode, the metal ion in the solution will accept one or more electrons from the cathode, and the ion's oxidation state will reduce to 0. This forms a solid metal that deposits on the cathode. The two electrodes must be electrically connected to each other, allowing for a flow of electrons that leave the metal of the anode and flow through this connection to the ions at the surface of the cathode. This flow of electrons is an electrical current that can be used to do work, such as turn a motor or power a light.

Example Reaction

The operating principle of the voltaic cell is a simultaneous oxidation and reduction reaction, called a redox reaction. This redox reaction consists of two half-reactions. In a typical voltaic cell, the redox pair is copper and zinc, represented in the following half-cell reactions:

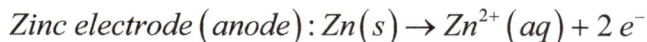

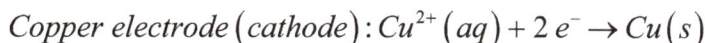

$$Zinc\ electrode\ (anode): Zn(s) \rightarrow Zn^{2+}(aq) + 2\ e^-$$

$$Copper\ electrode\ (cathode): Cu^{2+}(aq) + 2\ e^- \rightarrow Cu(s)$$

The cells are constructed in separate beakers. The metal electrodes are immersed in electrolyte solutions. Each half-cell is connected by a salt bridge, which allows for the free transport of ionic species between the two cells. When the circuit is complete, the current flows and the cell "produces" electrical energy.

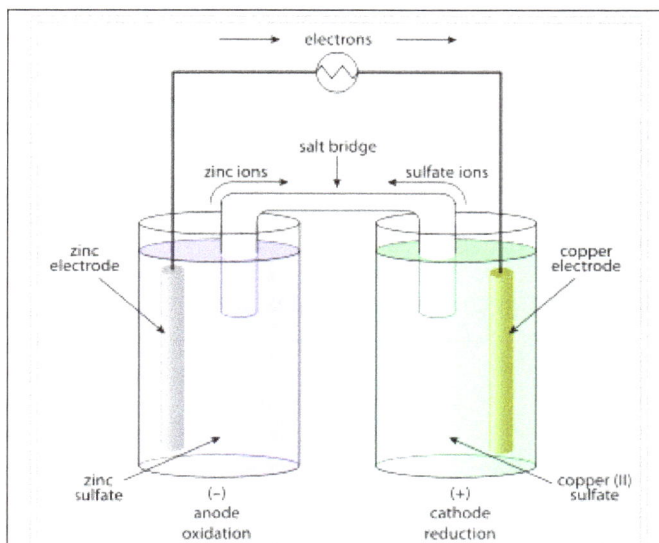

A galvanic, or voltaic, cell: The cell consists of two half-cells connected via a salt bridge or permeable membrane. The electrodes are immersed in electrolyte solutions and connected through an electrical load.

Copper readily oxidizes zinc; the anode is zinc and the cathode is copper. The anions in the solutions are sulfates of the respective metals. When an electrically conducting device connects the electrodes, the electrochemical reaction is:

$$Zn + Cu^2 + \rightarrow Zn^{2+} + Cu$$

The zinc electrode produces two electrons as it is oxidized ($Zn \rightarrow Zn^{2+} + 2e^-$), which travel through the wire to the copper cathode. The electrons then find the Cu^{2+} in solution and the copper is reduced to copper metal ($Cu^{2+} + 2e^- \rightarrow Cu$). During the reaction, the zinc electrode will be used and the metal will shrink in size, while the copper electrode will become larger due to the deposited Cu that is being produced. A salt bridge is necessary to keep the charge flowing through the cell. Without a salt bridge, the electrons produced at the anode would build up at the cathode and the reaction would stop running.

Voltaic cells are typically used as a source of electrical power. By their nature, they produce direct current. A battery is a set of voltaic cells that are connected in parallel. For instance, a lead–acid battery has cells with the anodes composed of lead and cathodes composed of lead dioxide.

Electrolytic Cells

Electrolysis uses electrical energy to induce a chemical reaction, which then takes place in an electrolytic cell.

In chemistry and manufacturing, electrolysis is a method of using a direct electric current (DC) to drive an otherwise non-spontaneous chemical reaction. Electrolysis is commercially important as a stage in the process of separating elements from naturally occurring sources such as ore.

Electrolysis is the passage of a direct electric current through an ionic substance that is either molten or dissolved in a suitable solvent, resulting in chemical reactions at the electrodes and separation of the materials.

Electrolysis can sometimes be thought of as running a non-spontaneous galvanic cell. Depending on how freely elements give up electrons (oxidation) and how energetically favorable it is for elements to receive electrons (reduction), the reaction may not be spontaneous. By externally supplying the energy to overcome the energy barrier to spontaneous reaction, the desired reaction is "allowed" to run under special circumstances.

The main components required to perform electrolysis are:

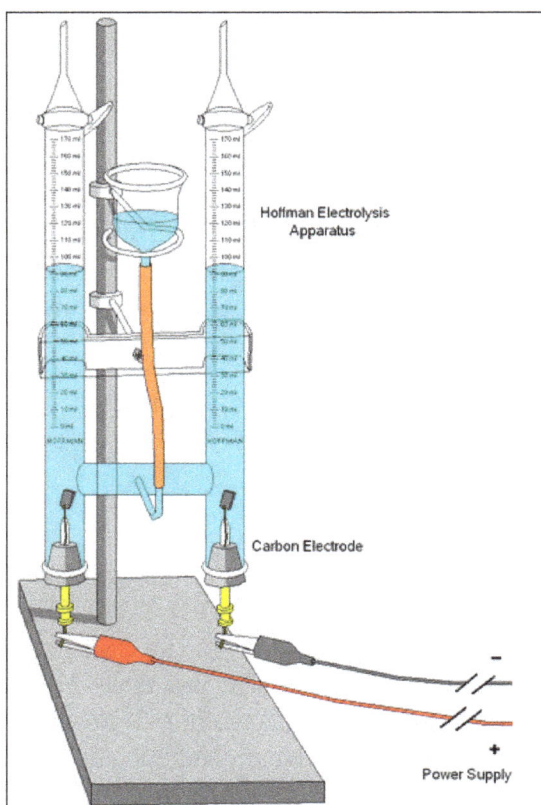

A typical electrolysis cell: A cell used in elementary chemical experiments to produce gas as a reaction product and to measure its volume.

- An electrolyte: A substance containing free ions that carry electric current. If the ions are not mobile, as in a solid salt, then electrolysis cannot occur.

- A direct current (DC) supply: Provides the energy necessary to create or discharge the ions in the electrolyte. Electric current is carried by electrons in the external circuit.

- Two electrodes: An electrical conductor that provides the physical interface between the electrical circuit providing the energy and the electrolyte.

Electrodes of metal, graphite, and semiconductor material are widely used. Choosing a suitable electrode depends on the chemical reactivity between the electrode and electrolyte, and the cost of manufacture.

Other systems that utilize the electrolytic process are used to produce metallic sodium and potassium, chlorine gas, sodium hydroxide, and potassium and sodium chlorate.

Electrochemical Cell Notation

Cell notation is shorthand that expresses a certain reaction in an electrochemical cell.

Cell Notation

Recall that standard cell potentials can be calculated from potentials E°_{cell} for both oxidation and reduction reactions. A positive cell potential indicates that the reaction proceeds spontaneously in the direction in which the reaction is written. Conversely, a reaction with a negative cell potential proceeds spontaneously in the reverse direction.

$$E^{\circ}_{cell} = E^{\circ}_{reduction} + E^{\circ}_{oxidation}$$

Cell notations are a shorthand description of voltaic or galvanic (spontaneous) cells. The reaction conditions (pressure, temperature, concentration, etc.), the anode, the cathode, and the electrode components are all described in this unique shorthand.

Recall that oxidation takes place at the anode and reduction takes place at the cathode. When the anode and cathode are connected by a wire, electrons flow from anode to cathode.

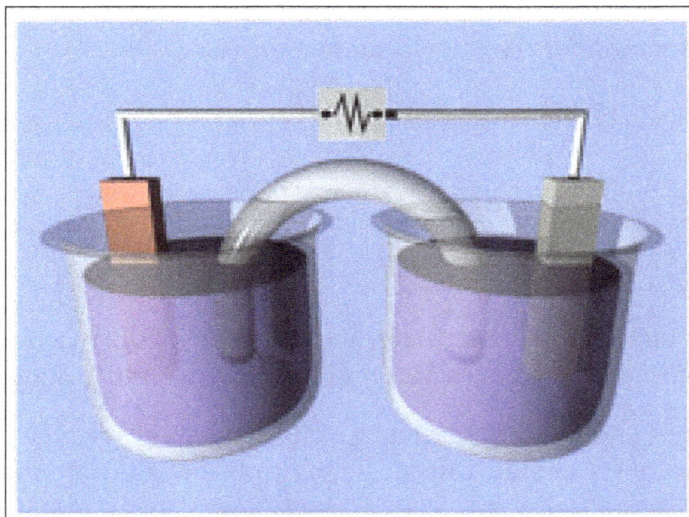

A typical galvanic cell: A typical arrangement of half-cells linked to form a galvanic cell.

Using the arrangement of components, let's put a cell together.

One beaker contains 0.15 M $Cd(NO_3)_2$ and a Cd metal electrode. The other beaker contains 0.20 M $AgNO_3$ and a Ag metal electrode. The net ionic equation for the reaction is written:

$$2Ag^+(aq) + Cd(s) \rightleftharpoons Cd^{2+}(aq) + 2Ag(s)$$

In the reaction, the silver ion is reduced by gaining an electron, and solid Ag is the cathode. The cadmium is oxidized by losing electrons, and solid Cd is the anode.

The anode reaction is:

$$Cd(s) \rightleftharpoons Cd^{2+}(aq) + 2e^-$$

The cathode reaction is:

$$2Ag^+(aq) + 2e^- \rightleftharpoons 2Ag(s)$$

Cell Notation Rules

1. The anode half-cell is described first; the cathode half-cell follows. Within a given half-cell, the reactants are specified first and the products last. The description of the oxidation reaction is first, and the reduction reaction is last; when you read it, your eyes move in the direction of electron flow. Spectator ions are not included.

2. A single vertical line $(|)$ is drawn between two chemical species that are in different phases but in physical contact with each other (e.g., solid electrode | liquid with electrolyte). A double vertical line $(||)$ represents a salt bridge or porous membrane separating the individual half-cells.

3. The phase of each chemical (s, l, g, aq) is shown in parentheses. If the electrolytes in the cells are not at standard conditions, concentrations and pressure, they are included in parentheses with the phase notation. If no concentration or pressure is noted, the electrolytes in the cells are assumed to be at standard conditions (1.00 M or 1.00 atm and 298 K).

Using these rules, the notation for the cell we put together is:

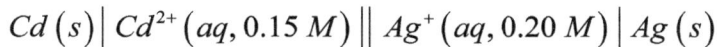

$$Cd(s) \mid Cd^{2+}(aq, 0.15\,M) \parallel Ag^+(aq, 0.20\,M) \mid Ag(s)$$

NERNST EQUATION

The Nernst Equation is the equation that relates the logarithm of the reaction quotient (Q) to non-standard cell potentials; can be used to relate equilibrium constants to standard cell potentials.

In galvanic cells, chemical energy is converted into electrical energy, which can do work. The electrical work is the product of the charge transferred multiplied by the potential difference (voltage):

electrical work = volts × (charge in coulombs) = J

The charge on 1 mole of electrons is given by Faraday's constant (F):

$$F = \frac{6.022 \times 10^{23}\,e^-}{mol} \times \frac{1.602 \times 10^{-19}\,C}{e^-} = 9.648 \times 10^4\,\frac{C}{mol} = 9.684 \times 10^4\,\frac{J}{V \cdot mol}$$

$$\text{total charge} = (\text{number of moles of } e^-) \times F = nF$$

In this equation, n is the number of moles of electrons for the balanced oxidation-reduction reaction. The measured cell potential is the maximum potential the cell can produce and is related to the electrical work (w_{ele}) by:

$$E_{cell} = \frac{-w_{ele}}{nF} \qquad \text{or} \qquad w_{ele} = -nFE_{cell}$$

The negative sign for the work indicates that the electrical work is done by the system (the galvanic cell) on the surroundings. The free energy was defined as the energy that was available to do work. In particular, the change in free energy was defined in terms of the maximum work (w_{max}), which, for electrochemical systems, is w_{ele}.

$$\Delta G = w_{max} = w_{ele}$$
$$\Delta G = -nFE_{cell}$$

We can verify the signs are correct when we realize that n and F are positive constants and that galvanic cells, which have positive cell potentials, involve spontaneous reactions. Thus, spontaneous reactions, which have $\Delta G < 0$, must have $E_{cell} > 0$. If all the reactants and products are in their standard states, this becomes:

$$\Delta G^\circ = -nFE_{cell}^\circ$$

This provides a way to relate standard cell potentials to equilibrium constants, since:

$$\Delta G^\circ = -RT \ln K$$

$$-nFE_{cell}^\circ = -RT \ln K \qquad or \qquad E_{cell}^\circ = \frac{RT}{nF} \ln K$$

Most of the time, the electrochemical reactions are run at standard temperature (298.15 K). Collecting terms at this temperature yields:

$$E_{cell}^\circ = \frac{RT}{nF} \ln K = \frac{(8.314\,\frac{J}{K \times mol})(298.15K)}{n \times 96,485\,C/V \times mol} \ln K = \frac{0.0257\,V}{n} \ln K$$

where n is the number of moles of electrons. For historical reasons, the logarithm in equations involving cell potentials is often expressed using base 10 logarithms (log), which changes the constant by a factor of 2.303:

$$E_{cell}^\circ = \frac{0.0592\,V}{n} \log K$$

Thus, if $\Delta G°$, K, or $E°_{cell}$ is known or can be calculated, the other two quantities can be readily determined. The relationships are shown graphically in the figure.

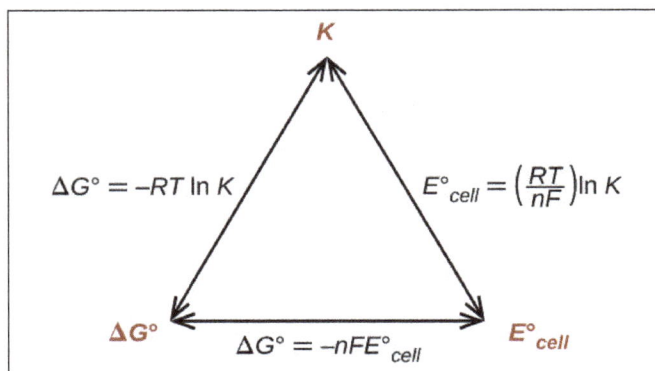

The relationships between $\Delta G°$, K, and $E°_{cell}$. Given any one of the three quantities, the other two can be calculated, so any of the quantities could be used to determine whether a process was spontaneous.

Given any one of the quantities, the other two can be calculated.

Now that the connection has been made between the free energy and cell potentials, nonstandard concentrations follow. Recall that:

$$\Delta G = \Delta G° + RT \ln Q$$

where Q is the reaction quotient. Converting to cell potentials:

$$-nFE_{cell} = -nFE°_{cell} + RT \ln Q \qquad or \qquad E_{cell} = E°_{cell} - \frac{RT}{nF} \ln Q$$

This is the Nernst equation. At standard temperature (298.15 K), it is possible to write the above equations as:

$$E_{cell} = E°_{cell} - \frac{0.0257\,V}{n} \ln Q \qquad or \qquad E_{cell} = E°_{cell} - \frac{0.0592\,V}{n} \log Q$$

If the temperature is not 273.15 K, it is necessary to recalculate the value of the constant. With the Nernst equation, it is possible to calculate the cell potential at nonstandard conditions. This adjustment is necessary because potentials determined under different conditions will have different values.

EQUILIBRIUM CONSTANT

We need to look at two different types of equilibria (homogeneous and heterogeneous) separately, because the equilibrium constants are defined differently.

- A homogeneous equilibrium has everything present in the same phase. The usual examples include reactions where everything is a gas, or everything is present in the same solution.

- A heterogeneous equilibrium has things present in more than one phase. The usual examples include reactions involving solids and gases, or solids and liquids.

K_c In Homogeneous Equilibria

This is the more straightforward case. It applies where everything in the equilibrium mixture is present as a gas, or everything is present in the same solution.

A good example of a gaseous homogeneous equilibrium is the conversion of sulphur dioxide to sulphur trioxide at the heart of the Contact Process:

$$2SO_{2(g)} + O_{2(g)} \rightleftharpoons 2SO_{2(g)}$$

A commonly used liquid example is the esterification reaction between an organic acid and an alcohol - for example:

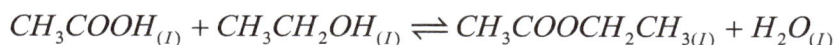

$$CH_3COOH_{(l)} + CH_3CH_2OH_{(l)} \rightleftharpoons CH_3COOCH_2CH_{3(l)} + H_2O_{(l)}$$

Writing an Expression for K_c

We are going to look at a general case with the equation:

$$aA + bB + \rightleftharpoons cC + dD$$

No state symbols have been given, but they will be all (g), or all (l), or all (aq) if the reaction was between substances in solution in water.

If you allow this reaction to reach equilibrium and then measure the equilibrium concentrations of everything, you can combine these concentrations into an expression known as an equilibrium constant.

The equilibrium constant always has the same value (provided you don't change the temperature), irrespective of the amounts of A, B, C and D you started with. It is also unaffected by a change in pressure or whether or not you are using a catalyst.

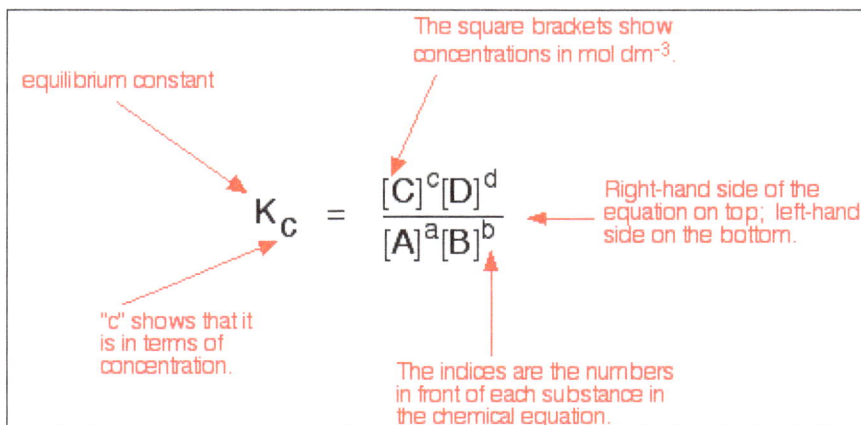

equilibrium constant

The square brackets show concentrations in mol dm^{-3}.

$$K_c = \frac{[C]^c[D]^d}{[A]^a[B]^b}$$

Right-hand side of the equation on top; left-hand side on the bottom.

"c" shows that it is in terms of concentration.

The indices are the numbers in front of each substance in the chemical equation.

Compare this with the chemical equation for the equilibrium. The convention is that the substances on the right-hand side of the equation are written at the top of the K_c expression, and those on the left-hand side at the bottom.

The indices (the powers that you have to raise the concentrations to - for example, squared or cubed or whatever) are just the numbers that appear in the equation.

Some Specific Examples

The Esterification Reaction Equilibrium

A typical equation might be:

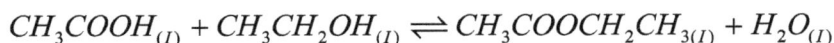

$$CH_3COOH_{(l)} + CH_3CH_2OH_{(l)} \rightleftharpoons CH_3COOCH_2CH_{3(l)} + H_2O_{(l)}$$

There is only one molecule of everything shown in the equation. That means that all the powers in the equilibrium constant expression are "1". You don't need to write those into the K_c expression.

$$K_c = \frac{[CH_3COOCH_2CH_3][H_2O]}{[CH_3COOH][CH_3CH_2OH]}$$

As long as you keep the temperature the same, whatever proportions of acid and alcohol you mix together, once equilibrium is reached, K_c always has the same value. At room temperature, this value is approximately 4 for this reaction.

The Equilibrium in the Hydrolysis of Esters

This is the reverse of the last reaction:

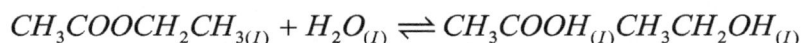

$$CH_3COOCH_2CH_{3(l)} + H_2O_{(l)} \rightleftharpoons CH_3COOH_{(l)}CH_3CH_2OH_{(l)}$$

The K_c expression is:

$$K_c = \frac{[CH_3COOH][CH_3CH_2OH]}{[CH_3COOCH_2CH_3][H_2O]}$$

If you compare this with the previous example, you will see that all that has happened is that the expression has turned upside-down. Its value at room temperature will be approximately 1/4 (0.25).

It is really important to write down the equilibrium reaction whenever you talk about an equilibrium constant. That is the only way that you can be sure that you have got the expression the right way up - with the right-hand substances on the top and the left-hand ones at the bottom.

The Contact Process Equilibrium

You will remember that the equation for this is:

$$2SO_{2(g)} + O_{2(g)} \rightleftharpoons 2SO_{3(g)}$$

This time the K_c expression will include some visible powers:

$$K_c = \frac{[SO_3]^2}{[SO_2]^2 [O_2]}$$

Although everything is present as a gas, you still measure concentrations in mol dm^{-3}. There is another equilibrium constant called K_p which is more frequently used for gases.

The Haber Process Equilibrium

The equation for this is:

$$N_{2(g)} + 3H_{2(g)} \rightleftharpoons 2NH_{3(g)}$$

and the K_c expression is:

$$K_c = \frac{[NH_3]^2}{[N_2][H_2]^3}$$

K_c in Heterogeneous Equilibria

Typical examples of a heterogeneous equilibrium include:

The equilibrium established if steam is in contact with red hot carbon. Here we have gases in contact with a solid.

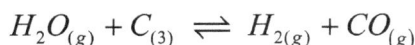

$$H_2O_{(g)} + C_{(3)} \rightleftharpoons H_{2(g)} + CO_{(g)}$$

If you shake copper with silver nitrate solution, you get this equilibrium involving solids and aqueous ions:

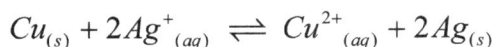

$$Cu_{(s)} + 2Ag^+{}_{(aq)} \rightleftharpoons Cu^{2+}{}_{(aq)} + 2Ag_{(s)}$$

Writing an Expression for K_c for a Heterogeneous Equilibrium

The important difference this time is that you don't include any term for a solid in the equilibrium expression.

Taking another look at the two examples above, and adding a third one:

The Equilibrium Produced on Heating Carbon with Steam

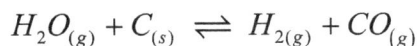

$$H_2O_{(g)} + C_{(s)} \rightleftharpoons H_{2(g)} + CO_{(g)}$$

Everything is exactly the same as before in the equilibrium constant expression, except that you leave out the solid carbon.

$$K_c = \frac{[H_2][CO]}{[H_2O]}$$

The Equilibrium Produced between Copper and Silver Ions

$$Cu_{(s)} + 2Ag^+_{(aq)} \rightleftharpoons Cu^{2+}_{(aq)} + 2Ag_{(s)}$$

Both the copper on the left-hand side and the silver on the right are solids. Both are left out of the equilibrium constant expression.

$$K_c = \frac{[Cu^{2+}]}{[Ag^+]^2}$$

The Equilibrium Produced on Heating Calcium Carbonate

This equilibrium is only established if the calcium carbonate is heated in a closed system, preventing the carbon dioxide from escaping.

$$CaCO_{3(s)} \rightleftharpoons CaO(s) + CO_{2(g)}$$

The only thing in this equilibrium which isn't a solid is the carbon dioxide. That is all that is left in the equilibrium constant expression.

$$K_c = [CO_2]$$

Calculations involving K_c

There are all sorts of calculations you might be expected to do which are centred around equilibrium constants. You might be expected to calculate a value for K_c including its units (which vary from case to case). Alternatively you might have to calculate equilibrium concentrations from a given value of K_c and given starting concentrations.

MASS TRANSFER

Mass transfer describes the transport of mass from one point to another and is one of the main pillars in the subject of Transport Phenomena. Mass transfer may take place in a single phase or over phase boundaries in multiphase systems. In the vast majority of engineering problems, mass transfer involves at least one fluid phase (gas or liquid), although it may also be described in solid-phase materials.

In many cases, the mass transfer of species takes place together with chemical reactions. This implies that flux of a chemical species does not have to be conserved in a volume element, since chemical species may be produced or consumed in such an element. The chemical reactions are sources or sinks in such flux balances.

The theory of mass transfer allows for the computation of mass flux in a system and the distribution of the mass of different species over time and space in such a system, also when chemical reactions are present. The purpose of such computations is to understand, and possibly design or control, such a system.

Mathematical Description of Mass Transfer

The driving force, F, for mass transfer is created by gradients in the system potential, U:

$$F = -\nabla U$$

Gradients in chemical composition are usually responsible for this driving force. The driving force for transport over phase boundaries is generated by a deviation from equilibrium over such a phase boundary. Additional driving forces may contribute with a drift velocity, such as the forces created by migration, pressure, gravitational, and centrifugal forces.

The equation below shows the forces acting on a chemical species, per mole of atoms, ions, or molecules, due to gradients in chemical potential and electric fields (migration).

$$F_i = -\underbrace{RT\nabla \ln a_i}_{\text{Chemical}} - \underbrace{z_i F \nabla \phi}_{\text{Electrical}}$$

In these equations, R denotes the gas constant, T is the temperature, a_i is the activity of each species, z_i denotes the charge number of a species, F is the Faraday constant, and ϕ is the electric potential. The negative gradient of ϕ is the electric field. The activity can be understood as a thermodynamic measure of the chemical potential of the system, so that gradients in activity correspond to driving forces for chemical mass transport.

A simple chemical assumption is that the activity of a species i is given by its mole fraction, denoted as x_i. This is exactly true for an ideal mixture:

$$F_i = -RT\frac{1}{x_i}\nabla x_i - z_i F \nabla \phi$$

The forces on a species i are balanced by the friction in the interaction between this species and the other species in a mixture. The friction acting on a mole of i is proportional to the difference in mass velocity between i and each species j in the mixture, the mole fraction of each species j in a mixture, and the friction coefficient between i and j.

$$F_{\text{fric},i} = -\sum_{j \neq i} \zeta_{ij} x_j (u_{R,i} - u_{R,j})$$

In this equation, ζ_{ij} denotes the friction coefficient between species i and j, x_j is the mole fraction of species j, and $u_{R,i}$ is the mass species velocity of species i relative to the mass average velocity of the whole mixture. Note that the mass velocities of each species in the equation above are given using the mass average velocity for the mixture as reference. A species that does not deviate from the velocity of the mixture (i.e., that does not diffuse or migrate, in this case), has a zero $u_{R,i}$, when using the mixture velocity as reference.

If we now set the driving forces to exactly balance the friction forces acting on species i, we obtain the following equation:

$$-RT\frac{1}{x_i}\nabla x_i - z_i F \nabla \phi - \sum_{j \neq i} \zeta_{ij} x_j (u_{R,i} - u_{R,j}) = 0$$

The molar flux is defined as:

$$J_i = c x_i u_{R,i}$$

where J_i is the flux vector of species i relative to the velocity of the mixture and c is the total concentration of all species in a mixture. Introducing the Maxwell-Stefan diffusivity as:

$$Đ_{ij} = \frac{1}{\zeta_{ij} RT}$$

and using the molar flux to eliminate $u_{R,i}$ and $u_{R,j}$ in the force balance equation above yields the following expression:

$$- c \nabla x_i - \frac{z_i Fc}{RT} x_i \nabla \phi = \sum_{j \neq i} \frac{1}{Đ_{ij}} (x_i J_i - x_i J_i)$$

This is the Maxwell-Stefan equation, an equation that forms the basis for the mathematical description of mass transfer of chemical species in a mixture. Simplifications of these equations for diluted mixtures give, for example, Fick's first law of diffusion and the Nernst-Planck equations for diffusion and migration.

The molar flux of a species i relative to a fixed coordinate system, denoted as N_i, is obtained by adding the convective term, due to the velocity of the whole mixture:

$$N_i - J_i + c_i u$$

The resulting fluxes are used in the mass conservation equations for each species in the solution:

$$\frac{\partial M_i c_i}{\partial t} + \nabla \cdot M_i N_i - M_i R_i = 0$$

The sum of all mass fluxes, including the convective term, results in the continuity equation for the mixture:

$$\frac{\partial}{\partial t} \underbrace{\sum_i M_i c_i}_{= \rho} + \nabla \cdot \underbrace{\sum_i M_i N_i}_{= \rho u} - \underbrace{\sum_i M_i R_i}_{= 0} = 0$$

where the last term is necessarily zero due to the conservation of mass of an individual chemical reaction. By identifying the sums as the density and mass flux density, we get the mass continuity equation:

$$\frac{\partial}{\partial t} + \nabla \cdot \rho u = 0$$

The convective term in the flux is the contribution to the flux of a species due to the movement of the whole solution. For this reason, convective flux takes place along the velocity streamlines of the solution for all chemical species in the solution. Note that the sum of all the species' mass fluxes, relative to the flux of the mixture, is zero if the mass-averaged velocity is used as reference. The mass-averaged velocity is defined as:

$$u = \frac{\sum_i \rho_i u_i}{\sum_i \rho_i}$$

where ∂_i denotes the mass density of a species i. This implies that, in general, the mass flux of each species is tightly coupled to the total mass velocity in a mixture. In a strict definition, the mass average velocity of a mixture could be obtained by formulating and solving the equations for the conservation of momentum for each species in a mixture.

However, the interaction coefficients, required for such formulations, are usually difficult to measure or calculate. Instead, equations for the conservation of momentum for the whole mixture are usually defined. The combination of the equations for the conservation of momentum and mass for a mixture at low velocities (less than one third of the speed of sound), yield the Navier-Stokes equations. The solution of the Navier-Stokes equations gives the velocity field (a vector field) that also determines the direction of the convective flux of all species in the mixture.

The tight coupling between mass transport for each species and the conservation of mass for the whole mixture is exemplified in the example below. Oxygen in air is consumed at the surface of a catalyst and produces liquid water that is removed from the gas phase in a gas diffusion electrode. The consumption of oxygen causes a net velocity in the gas mixture (air). Additionally, a nitrogen concentration gradient is formed in order to perfectly balance the advective (or convective) flux of nitrogen with an opposite flux by diffusion.

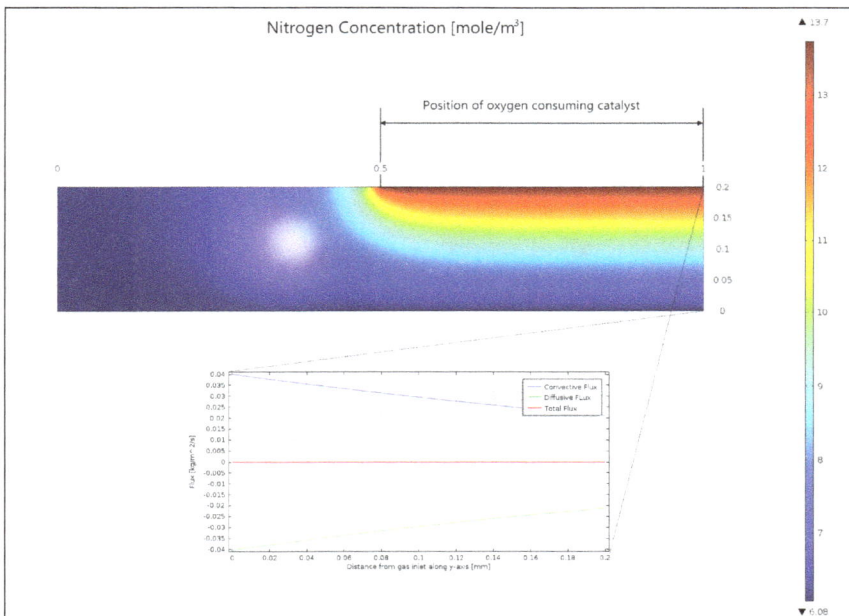

Surface plot of the nitrogen concentration in a gas diffusion electrode. The flux along the right vertical edge, plotted in the x-y plot, shows that the diffusive flux from the catalytic surface exactly compensates the convective flux to this surface, generated by oxygen consumption at the catalyst's surface.

Due to the difficulties of numerically resolving steep gradients in potential, mass transfer across phase boundaries is often expressed using difference equations, instead of differential equations. This approximation implies that the gradients, included in the driving forces, are linearized inside a fictitious boundary layer. The thickness of the layer is then defined as the distance from a phase boundary where the linearized concentration gradient, starting from the concentration at the phase boundary, reaches the bulk concentration. The definition of the boundary layer also means that its thickness may be different for different species.

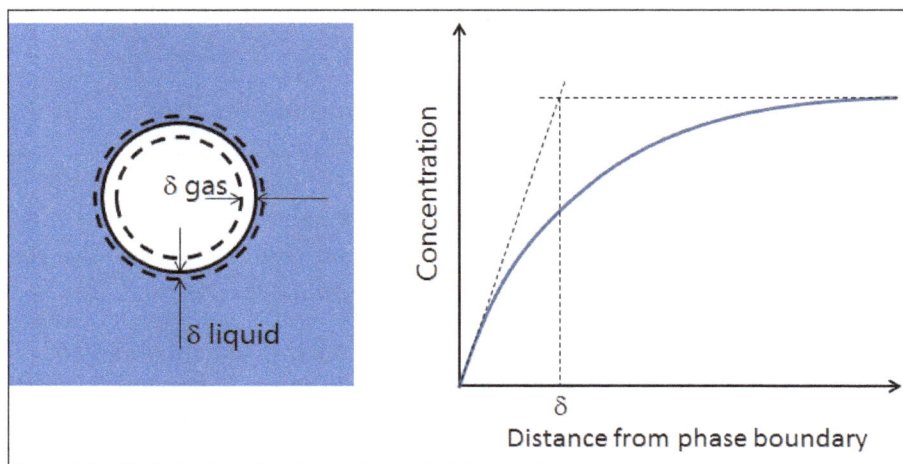

Fictive boundary layers inside and around a gas bubble in a liquid.

The mass transfer coefficient, k_m, for such an interface is defined as the diffusivity divided by the boundary layer thickness, δ.

$$k_m = \frac{D}{\delta}$$

An estimate of the relation between the boundary layer thickness and a system's typical length is given by the Sherwood number:

$$Sh = \frac{k_m L}{D}$$

In this expression, L denotes a typical length for a system, such as the radius of a pipe and the width of a channel. However, if the mass transport around a gas bubble in a liquid is studied, then L may denote the radius of the bubble. Since the thickness of the boundary layer depends on the convection just outside an interface, the Sherwood number also gives a measurement of the convective and diffusive fluxes to such an interface.

The boundary layer thickness at a liquid-gas interface for a rising bubble is of the order of magnitude of 100 μm in the gas phase and around 10 μm in the liquid phase.

The Sherwood number can also be defined as a function of the Reynolds and Schmidt numbers. The Reynolds number gives an estimate of the ratio of momentum transport by inertia to viscosity in a fluid:

$$R_e = \frac{\rho U L}{\mu}$$

where μ denotes viscosity and U denotes an average velocity. The Schmidt number gives an estimate of the relation between viscosity and diffusivity in a fluid:

$$Sc = \frac{\mu}{\rho D}$$

The mass transfer coefficient can be estimated from the relation between the Sherwood number and the Reynolds and Schmidt numbers. For example, for forced convection along a flat plate, the following expression can be used:

$$\mathrm{Sh} = 0.5\, f_{\mathrm{loc}}\, \mathrm{Re}\, \mathrm{Sc}^{\frac{1}{3}}$$

Where f_{loc} denotes the local friction factor for flow along a flat plate. The friction coefficient for different geometries is tabulated in literature and may also be obtained experimentally. All the material properties and the average velocity in the relation above are relatively easy to find in literature or estimate from simple calculations. Once the Sherwood number is calculated, then the mass transfer coefficient can be calculated, including the boundary layer thickness, which is the parameter that is not easily estimated otherwise.

COTTRELL EQUATION

The Cottrell equation in electrochemistry gives the current to a planar electrode in the potential-step experiment. This Demonstration shows the dependence of the current on the time, concentration, and diffusion coefficient.

The Cottrell equation is:

$$I = \frac{nF\, A\sqrt{D}\, c}{\sqrt{\pi t}},$$

where D is the diffusion coefficient (m^2/s), c is the concentration in bulk solution (mM), A is the surface area of the electrode (m^2), F is Faraday's constant, t is the time (s), and n is the number of electrons transferred.

The graphic on the left shows that the current of the electrode is inversely proportional to $\sqrt{\pi t}$; the current decays quickly for a short time and tends to a limiting value with increasing time.

This phenomenon can be explained by the concentration profile (right), since the current is related to $\partial_x c(x,t)\, \delta = \sqrt{\pi D t}$ at $x = 0$. Initially, $\partial_x c(x,t)$ is very large, so the current is large. As time increases, the concentration near the electrode surface is consumed and $\partial_x c(x,t)$ decreases. The filling part of the concentration profile represents the Nernst diffusion layer $\delta = \sqrt{\pi D t}$, which is the distance when the concentration consumed is about 80%.

ELECTRODE POTENTIAL

Standard electrode potential is a measurement of the potential for equilibrium. There is a potential difference between the electrode and the electrolyte called the potential of the electrode. When unity is the concentrations of all the species involved in a semi-cell, the electrode potential is known as the standard electrode potential.

Standard Electrode Potential

Under standard conditions, the standard electrode potential occurs in an electrochemical cell say the temperature = 298K, pressure = 1atm, concentration = 1M. The symbol 'E°_{cell}' represents the standard electrode potential of a cell.

Significance of Standard Electrode Potential

- All electrochemical cells are based on redox reactions, which are made up of two half-reactions.

- The oxidation half-reaction occurs at the anode and it involves a loss of electrons.

- Reduction reaction takes place at the cathode, involving a gain of electrons. Thus, the electrons flow from the anode to the cathode.

- The electric potential that arises between the anode and the cathode is due to the difference in the individual potentials of each electrode (which are dipped in their respective electrolytes).

- The cell potential of an electrochemical cell can be measured with the help of a voltmeter. However, the individual potential of a half-cell cannot be accurately measured alone.

- It is also important to note that this potential can vary with a change in pressure, temperature, or concentration.

- In order to obtain the individual reduction potential of a half-cell, the need for standard electrode potential arises.

- It is measured with the help of a reference electrode known as the standard hydrogen electrode (abbreviated to SHE). The electrode potential of SHE is zero Volts.

- The standard electrode potential of an electrode can be measured by pairing it with the SHE and measuring the cell potential of the resulting galvanic cell.

- The oxidation potential of an electrode is the negative of its reduction potential. Therefore, the standard electrode potential of an electrode is described by its standard reduction potential.

- Good oxidizing agents have high standard reduction potentials whereas good reducing agents have low standard reduction potentials.

- For example, the standard electrode potential of Ca^+ is -3.8V and that of F_2 is +2.87V. This implies that F_2 is a good oxidizing agent whereas Ca is a reducing agent.

Standard Electrode Potential Example

The calculation of the standard electrode potential of a zinc electrode with the help of the standard hydrogen electrode is illustrated below.

It can be noted that this potential is measured under standard conditions where the temperature is 298K, the pressure is 1 atm, and the concentration of the electrolytes is 1M.

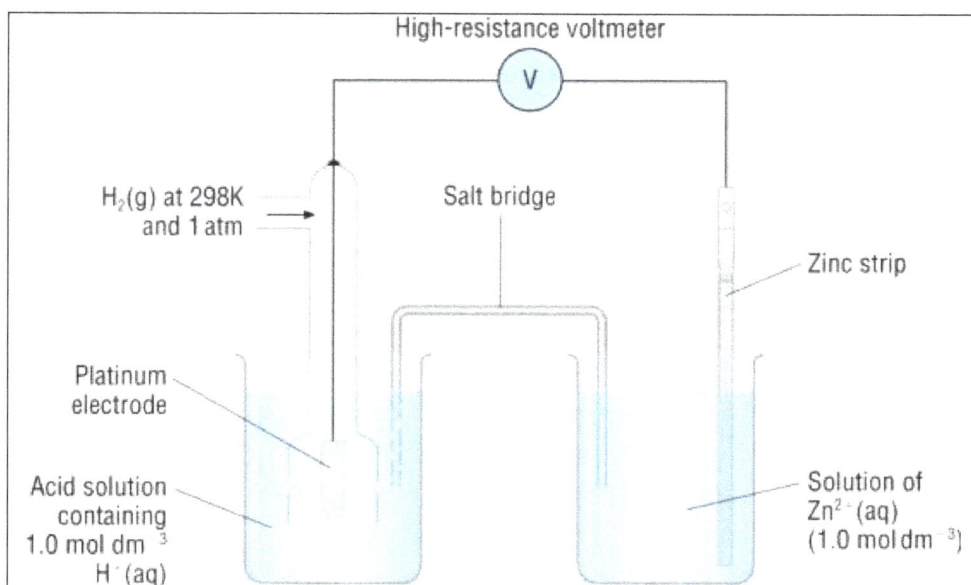

Spontaneity of Redox Reactions

If a redox reaction is spontaneous, the $\Delta G°$ (Gibbs free energy) must have a negative value. It is described by the following equation:

$$\Delta G°_{cell} = -nFE°_{cell}$$

Where n refers to the total number of moles of electrons for every mole of product formed, F is Faraday's constant (approximately 96485 C.mol^{-1}).

The $E°_{cell}$ can be obtained with the help of the following equation:

$$E°_{cell} = E°_{cathode} - E°_{anode}$$

Therefore, the $E°_{cell}$ can be obtained by subtracting the standard electrode potential of the anode from that of the cathode. For a redox reaction to be spontaneous, the $E°_{cell}$ must have a positive value (because both n and F have positive positive values, and the $\Delta G°$ value must be negative).

This implies that in a spontaneous process,

$$E°_{cell} > 0; \text{ which in turn implies that } E°_{cathode} > E°_{anode}$$

Thus, the standard electrode potential of the cathode and the anode help in predicting the spontaneity of the cell reaction. It can be noted that the $\Delta G°$ of the cell is negative in galvanic cells and positive in electrolytic cells.

References

- Electrochemistry, Di-Fa: chemistryexplained.com, Retrieved 31 March, 2019
- Photoelectrochemistry, 4813, definition: corrosionpedia.com, Retrieved 14 July, 2019

- Quantum-electrochemistry, chemistry, science: assignmentpoint.com, Retrieved 17 May, 2019

- Electrochemical-cell, chemistry: byjus.com, Retrieved 19 April, 2019

- Electrochemical-cells, chapter, boundless-chemistry: lumenlearning.com, Retrieved 25 February, 2019

- 17-4-the-nernst-equation, chemistry: opentextbc.ca, Retrieved 26 July, 2019

- Equilibria, physical: chemguide.co.uk, Retrieved 21 May, 2019

- What-is-mass-transfer, multiphysics: comsol.co.in, Retrieved 8 January, 2019

- Cottrell equation for the potentialst epexperiment: wolfram.com, Retrieved 13 May, 2019

- Standard-electrode-potential, chemistry: byjus.com, Retrieved 25 February, 2019

Electrochemical Reactions

The processes in which electrons flow between solid electrodes and a substance, known as electrolyte, are known as electrochemical reactions. Some of the examples of reactions are redox reactions, corrosion reaction and combination reaction. The chapter closely examines these key electrochemical reactions to provide an extensive understanding of the subject.

Electrochemical reaction is any process either caused or accompanied by the passage of an electric current and involving in most cases the transfer of electrons between two substances—one a solid and the other a liquid.

Under ordinary conditions, the occurrence of a chemical reaction is accompanied by the liberation or absorption of heat and not of any other form of energy; but there are many chemical reactions that—when allowed to proceed in contact with two electronic conductors, separated by conducting wires—liberate what is called electrical energy, and an electric current is generated. Conversely, the energy of an electric current can be used to bring about many chemical reactions that do not occur spontaneously. A process involving the direct conversion of chemical energy when suitably organized constitutes an electrical cell. A process whereby electrical energy is converted directly into chemical energy is one of electrolysis; i.e., an electrolytic process. By virtue of their combined chemical energy, the products of an electrolytic process have a tendency to react spontaneously with one another, reproducing the substances that were reactants and were therefore consumed during the electrolysis. If this reverse reaction is allowed to occur under proper conditions, a large proportion of the electrical energy used in the electrolysis may be regenerated. This possibility is made use of in accumulators or storage cells, sets of which are known as storage batteries. The charging of an accumulator is a process of electrolysis; a chemical change is produced by the electric current passing through it. In the discharge of the cell, the reverse chemical change occurs, the accumulator acting as a cell that produces an electric current.

Finally, the passage of electricity through gases generally causes chemical changes, and this kind of reaction forms a separate branch of electrochemistry that will not be treated here.

General Principles

Substances that are reasonably good conductors of electricity may be divided into two groups: the metallic, or electronic, conductors and the electrolytic conductors. The metals and many nonmetallic substances such as graphite, manganese dioxide, and lead sulfide exhibit metallic conductivity; the passage of an electric current through them produces heating and magnetic effects but no chemical changes. Electrolytic conductors, or electrolytes, comprise most acids, bases, and salts, either in the molten condition or in solution in water or other solvents. Plates or rods composed of a suitable metallic conductor dipping into the fluid electrolyte are employed to conduct the current into and out of the liquid; i.e., to act as electrodes. When a current is passed between electrodes through an electrolyte, not only are heating and magnetic effects produced but also definite chemical changes occur. At or in the neighbourhood of the negative electrode, called the cathode, the chemical change may be the deposition of a metal or the liberation of hydrogen and formation of a basic substance or some other chemical reduction process; at the positive electrode, or anode, it may be the dissolution of the anode itself, the liberation of a nonmetal, the production of oxygen and an acidic substance, or some other chemical oxidation process.

An electrolyte, prepared either by the melting of a suitable substance or by the dissolving of it in water or other liquid, owes its characteristic properties to the presence in it of electrically charged atoms or groups of atoms produced by the spontaneous splitting up or dissociation of the molecules of the substance. In solutions of the so-called strong electrolytes, most of the original substance, or in some solutions perhaps all of it, has undergone this process of electrolytic dissociation into charged particles, or ions. When an electrical potential difference (i.e., a difference in degree of electrification) is established between electrodes dipping into an electrolyte, positively charged ions move toward the cathode and ions bearing negative charges move toward the anode. The electric current is carried through the electrolyte by this migration of the ions. When an ion reaches the electrode of opposite polarity, its electrical charge is donated to the metal, or an electric charge is received from the metal. The ion is thereby converted into an ordinary neutral atom or group of atoms. It is this discharge of ions that gives rise to one of the types of chemical changes occurring at electrodes.

Sites of Electrochemical Reactions

Electrochemical reactions take place where the electron conductor meets the ionic conductor—i.e., at the electrode–electrolyte interface. Characteristic of this region, considered to be a surface phase, is the existence of a specific structure of particles and the presence of an electric field of considerable intensity (up to 10,000,000 volts per centimetre) across it; the field is caused by the separation of charges that are present between the two bulk phases in contact. For most purposes the surface phase can be considered as a parallel plate condenser, with one plate on the centre of the ions that have been brought to the electrode, at the distance of their closest approach to it, and with the second plate at the metal surface; between the two plates and acting as a dielectric (i.e., a nonconducting material) are oriented water molecules. This structure is termed the electric double layer and is illustrated in figure.

Thermal motion of the positive ions in the solution makes the condenser plate on the electrolyte side of the interface diffuse—i.e., the ions are distributed in a cloudlike way. This condition justifies the division of the potential change between the bulk of metal and the bulk of electrolyte into two parts: first, that between the metal surface and the first ionic layer at the distance of closest approach (called the outer Helmholtz plane, in which the ions are usually surrounded by solvent particles—i.e., are solvated); and second, that between the first ionic layer and the bulk of the solution, the diffuse part of the double layer. The picture is further complicated by the presence of ions in the electrode surface layer in addition to those that are present for electrostatic reasons—i.e., by the force of attraction or repulsion between electric charges. Such electrode surface layer ions are said to be specifically adsorbed on the electrode surface. Since this species of ions is attracted by the surface to a distance closer than the "distance of the closest approach" of ions, further subdivision of the inner part of the electric double layer is justified. Hence, the inner Helmholtz plane is introduced as the plane formed by the centres of specifically adsorbed ions. Adsorption of neutral molecules on the surface can also change the properties of the electric double layer. This change occurs as a consequence of replacing the water molecules, and thus changes that part of the potential (electrical) difference across the double layer that is caused by the adsorbed dipoles (water molecules that have a polarity—i.e., they behave like minute magnets—because of their hydrogen-oxygen structure, making one end of the molecule positive and the other end negative).

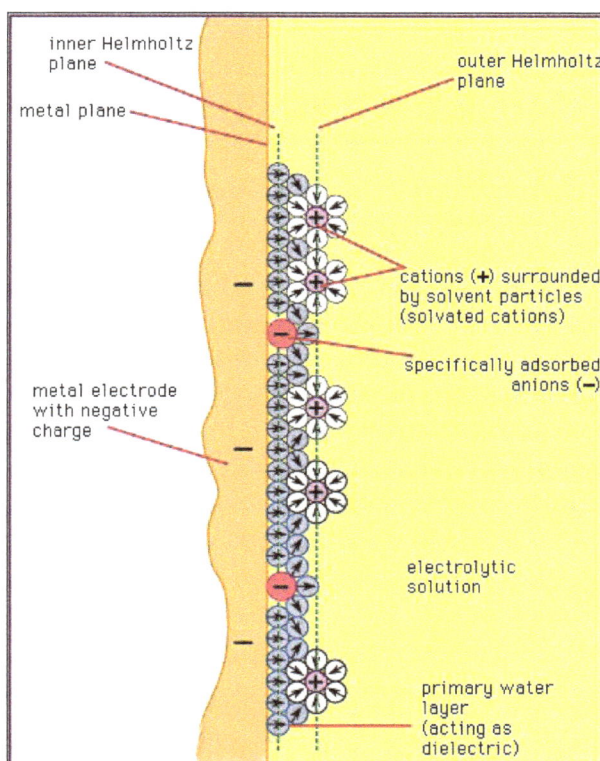

Double-layer structure and change of potential with distance from the electrode surface.

The absolute value of electrical potential difference, symbolized in calculation by the Greek letters delta and psi, $\Delta\psi$, between the bulk of a metal electrode and the bulk of an electrolyte cannot be measured. Instead, the voltage of a special cell, composed of the specific electrode being studied and of an arbitrarily selected reference electrode, is normally measured; the voltage is referred to as the relative electrode potential, (E) of special interest is that state of the electrode at which there is no net charge (in this case, no unbalanced, or extra positive, charge) at the metal side of the double layer. The relative potential at which this state is achieved is characteristic of each metal. This point is termed the potential of zero charge. At that potential, the field across the double layer is due to orientation of water molecules and other dipoles at the surface only.

Most of the knowledge of the detailed structure of the interface between a metal and an electrolyte arises from experimentation with mercury, the only metal that is liquid at ordinary temperatures; the double layer structure turns out to have surface tensions that must be measured, and this measurement is difficult with solid metals. By 1970, however, it had been shown that it is possible to measure surface tension changes at the metal-solution interface. Thus, the way to the determination of the double layer structure involving solids was opened.

Substances that are semiconductors can also be employed as electron carriers in electrochemical reactions. Semiconductors are substances which range between serving as insulators at low temperatures and as metallic-type conductors at high temperatures. In the case of semiconductors, however, the electric double layer has a more complex structure in as much as the condenser plate at the electrode side of the double layer also becomes diffuse. Thus, the overall potential difference between bulks of the phases in contact comprises also the potential difference between the bulk of the semiconductor and its surface.

ELECTROCHEMICAL REACTION MECHANISM

In chemistry, an electrochemical reaction mechanism is the step by step sequence of elementary steps, involving at least one outer sphere electron transfer, by which an overall chemical change occurs.

Elementary steps like proton coupled electron transfer and the movement of electrons between an electrode and substrate are special to electrochemical processes. Electrochemical mechanisms are important to all redox chemistry including corrosion, redox active photochemistry including photosynthesis, other biological systems often involving electron transport chains and other forms of homogeneous and heterogeneous electron transfer. Such reactions are most often studied with standard three electrode techniques such as cyclic voltammetry (CV), chronoamperometry, and bulk electrolysis as well as more complex experiments involving rotating disk electrodes and rotating ring-disk electrodes. In the case of photoinduced electron transfer the use of time-resolved spectroscopy is common.

Formalism

When describing electrochemical reactions an "E" and "C" formalism is often employed. The E represents an electron transfer; sometimes E_O and E_R are used to represent oxidations and reductions respectively. The C represents a chemical reaction which can be any elementary reaction step and is often called a "following" reaction. In coordination chemistry common C steps which "follow" electron transfer are ligand loss and association. The ligand loss or gain is associated with a geometric change in the complexes coordination sphere.

$$[ML_n]^{2+} + e^- \rightarrow [ML_n]^+ \quad E$$

$$[ML_n]^+ \rightarrow [ML_{(n-1)}]^+ + L \quad C$$

The reaction above would be called an EC reaction.

Characterization

The production of $[ML_{(n-1)}]^+$ in the reaction above by the "following" chemical reaction produces a species directly at the electrode that could display redox chemistry anywhere in a CV plot or none at all. The change in coordination from $[ML_n]^+$ to $[ML_{(n-1)}]^+$ often prevents the observation of "reversible" behavior during electrochemical experiments like cyclic voltammetry. On the forward scan the expected diffusion wave is observed, in example above the reduction of $[ML_n]^{2+}$ to $[ML_n]^{1+}$. However, on the return scan the corresponding wave is not observed, in the example above this would be the wave corresponding to the oxidation of $[ML_n]^{1+}$ to $[ML_n]^{2+}$. In our example there is no $[ML_n]^{1+}$ to oxidize since it has been converted to $[ML_{(n-1)}]^+$ through ligand loss. The return wave can sometimes be observed by increasing the scan rates so the following chemical reaction can be observed before the chemical reaction takes place. This often requires the use of ultramicroelectrodes (UME) capable of very high scan rates of 0.5 to 5.0 V/s. Plots of forward and reverse peak ratios against modified forms of the scan rate often identify the rate of the chemical reaction. It has become a common practice to model such plots with electrochemical simulations. The results of such studies are of disputed practical relevance since simulation requires excellent experimental data, better than that routinely obtained and reported. Furthermore, the parameters of such studies are rarely reported and often include an unreasonably high variable to data ratio. A better practice is to look for a simple, well documented relationship between observed results and implied phenomena; or to investigate a specific physical phenomenon using an alternative technique such as chrono-amperometry or those involving a rotating electrode.

Electrocatalysis

Electrocatalysis is a catalytic process involving oxidation or reduction through the direct transfer of electrons. The electrochemical mechanisms of electrocatalytic processes are a common research subject for various fields of chemistry and associated sciences. This is important to the development of water oxidation and fuel cells catalysts. For example, half the water oxidation reaction is the reduction of protons to hydrogen, the subsequent half reaction.

$$2H^+ + 2e^- \rightarrow H_2$$

This reaction requires some form of catalyst to avoid a large overpotential in the delivery of electrons. A catalyst can accomplish this reaction through different reaction pathways, two examples are listed below for the homogeneous catalysts $[ML_n]^{2+}$.

Pathway 1:

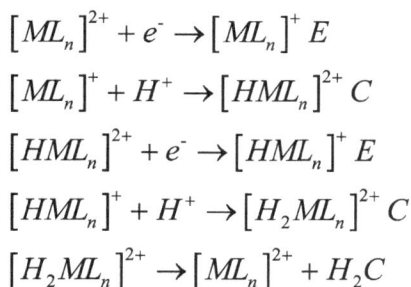

$$\left[ML_n\right]^{2+} + e^- \rightarrow \left[ML_n\right]^+ \quad E$$
$$\left[ML_n\right]^+ + H^+ \rightarrow \left[HML_n\right]^{2+} \quad C$$
$$\left[HML_n\right]^{2+} + e^- \rightarrow \left[HML_n\right]^+ \quad E$$
$$\left[HML_n\right]^+ + H^+ \rightarrow \left[H_2ML_n\right]^{2+} \quad C$$
$$\left[H_2ML_n\right]^{2+} \rightarrow \left[ML_n\right]^{2+} + H_2 \quad C$$

Pathway 2:

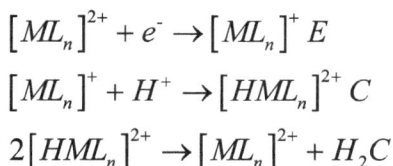

$$\left[ML_n\right]^{2+} + e^- \rightarrow \left[ML_n\right]^+ \quad E$$
$$\left[ML_n\right]^+ + H^+ \rightarrow \left[HML_n\right]^{2+} \quad C$$
$$2\left[HML_n\right]^{2+} \rightarrow \left[ML_n\right]^{2+} + H_2 \quad C$$

It is described as an ECECC while pathway 2 would be described as an ECC. If the catalyst was being considered for solid support pathway 1 which requires a single metal center to function would be a viable candidate. In contrast a solid support system which separates the individual metal centers would render a catalysts that operates through pathway 2 useless since it requires a step which is second order in metal center. Determining the reaction mechanism is much like other methods with some techniques unique to electrochemistry. In most cases electron transfer can be assumed to be much faster than the chemical reactions. Unlike stoichiometric reactions where the steps between the starting materials and the rate limiting step dominate in catalysis the observed reaction order is usually dominated by the steps between the catalytic resting state and the rate limiting step.

Following Physical Transformations

During potential variant experiments common to go through a redox couple in which the major species is transformed from a species that is soluble in the solution to one that is insoluble. This results in nucleation process in which a new species plates out on the working electrode. If a species has been deposited on the electrode during a potential sweep then on the return sweep a stripping wave is usually observed.

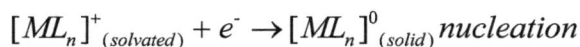

$$[ML_n]^+_{(solvated)} + e^- \rightarrow [ML_n]^0_{(solid)}\ nucleation$$

$$[ML_n]^0_{(solid)} \rightarrow e^- + [ML_n]^+_{(solvated)}\ stripping$$

While the nucleation wave may be pronounced or difficult the detect the stripping wave is usually very distinct. Often these phenomena can be avoided by reducing the concentration of the complex in solution. Neither these physical state changes involve a chemical reaction mechanism but they are worth mentioning here since the resulting data is at times confused with some chemical reaction mechanisms.

REDOX REACTION

Redox (short for reduction–oxidation reaction) is a type of chemical reaction in which the oxidation states of atoms are changed. Redox reactions are characterized by the transfer of electrons between chemical species, most often with one species (the reducing agent) undergoing oxidation (losing electrons) while another species (the oxidizing agent) undergoes reduction (gains electrons). The chemical species from which the electron is stripped is said to have been oxidized, while the chemical species to which the electron is added is said to have been reduced. In other words:

- Oxidation is the loss of electrons or an increase in the oxidation state of an atom by a molecule, an ion, or another atom.

- Reduction is the gain of electrons or a decrease in the oxidation state of an atom by a molecule, an ion, or another atom.

For example, during the combustion of wood, electrons are transferred from carbon atoms in the wood to oxygen atoms in the air. The oxygen atoms undergo reduction, gaining electrons, while the carbon atoms undergo oxidation, losing electrons. Thus oxygen is the oxidizing agent and carbon the reducing agent in this reaction.

Although oxidation reactions are commonly associated with the formation of oxides from oxygen molecules, oxygen is not necessarily included in such reactions, as other chemical species can serve the same function.

Redox reactions can occur relatively slowly, as in the formation of rust, or much more rapidly, as in the case of burning fuel. There are simple redox processes, such as the oxidation of carbon to yield carbon dioxide (CO_2) or the reduction of carbon by hydrogen to yield methane (CH_4), and more complex processes such as the oxidation of glucose ($C_6H_{12}O_6$) in the human body.

The processes of oxidation and reduction occur simultaneously and cannot happen independently of one another, similar to the acid–base reaction. The oxidation alone and the reduction alone are each called a half-reaction, because two half-reactions always occur together to form a whole reaction. When writing half-reactions, the gained or lost electrons are typically included explicitly in order that the half-reaction be balanced with respect to electric charge.

Though sufficient for many purposes, these general descriptions are not precisely correct. Although oxidation and reduction properly refer to a change in oxidation state — the actual transfer of electrons may never occur. The oxidation state of an atom is the fictitious charge that an atom would have if all bonds between atoms of different elements were 100% ionic. Thus, oxidation is best defined as an increase in oxidation state, and reduction as a decrease in oxidation state. In practice, the transfer of electrons will always cause a change in oxidation state, but there are many reactions that are classed as "redox" even though no electron transfer occurs (such as those involving covalent bonds).

Oxidizing and Reducing Agents

In redox processes, the reductant transfers electrons to the oxidant. Thus, in the reaction, the reductant or reducing agent loses electrons and is oxidized, and the oxidant or oxidizing agent gains electrons and is reduced. The pair of an oxidizing and reducing agent that are involved in a particular reaction is called a redox pair. A redox couple is a reducing species and its corresponding oxidizing form, e.g., Fe^{2+} / Fe^{3+}.

Oxidizers

The international pictogram for oxidizing chemicals.

Substances that have the ability to oxidize other substances (cause them to lose electrons) are said to be oxidative or oxidizing and are known as oxidizing agents, oxidants, or oxidizers. That is, the oxidant (oxidizing agent) removes electrons from another substance, and is thus itself reduced. And, because it "accepts" electrons, the oxidizing agent is also called an electron acceptor. Oxygen is the quintessential oxidizer.

Oxidants are usually chemical substances with elements in high oxidation states (e.g., H_2O_2, MnO_4^-, CrO_3, $Cr_2O_7^{2-}$, OsO_4), or else highly electronegative elements (O_2, F_2, Cl_2, Br_2) that can gain extra electrons by oxidizing another substance.

Reducers

Substances that have the ability to reduce other substances (cause them to gain electrons) are said to be reductive or reducing and are known as reducing agents, reductants, or reducers. The reductant (reducing agent) transfers electrons to another substance, and is thus itself oxidized. And, because it donates electrons, the reducing agent is also called an electron donor. Electron donors can also form charge transfer complexes with electron acceptors.

Reductants in chemistry are very diverse. Electropositive elemental metals, such as lithium, sodium, magnesium, iron, zinc, and aluminium, are good reducing agents. These metals donate or give away electrons readily. Hydride transfer reagents, such as $NaBH_4$ and $LiAlH_4$, are widely used in organic chemistry, primarily in the reduction of carbonyl compounds to alcohols. Another method of reduction involves the use of hydrogen gas (H_2) with a palladium, platinum, or nickel catalyst. These catalytic reductions are used primarily in the reduction of carbon-carbon double or triple bonds.

Standard Electrode Potentials (Reduction Potentials)

Each half-reaction has a standard electrode potential (E^0_{cell}), which is equal to the potential difference or voltage at equilibrium under standard conditions of an electrochemical cell in which the cathode reaction is the half-reaction considered, and the anode is a standard hydrogen electrode where hydrogen is oxidized:

$$^1/_2\, H_2 \rightarrow H^+ + e^-.$$

The electrode potential of each half-reaction is also known as its reduction potential E^0_{red}, or potential when the half-reaction takes place at a cathode. The reduction potential is a measure of the tendency of the oxidizing agent to be reduced. Its value is zero for $H^+ + e^- \rightarrow \frac{1}{2} H_2$ by definition, positive for oxidizing agents stronger than H+ (e.g., +2.866 V for F_2) and negative for oxidizing agents that are weaker than H+ (e.g., −0.763 V for Zn²⁺).

For a redox reaction that takes place in a cell, the potential difference is:

$$E^0_{cell} = E^0_{cathode} - E^0_{anode}$$

However, the potential of the reaction at the anode was sometimes expressed as an oxidation potential:

$$E^0_{ox} = -E^0_{red}.$$

The oxidation potential is a measure of the tendency of the reducing agent to be oxidized, but does not represent the physical potential at an electrode. With this notation, the cell voltage equation is written with a plus sign:

$$E^0_{cell} = E^0_{red(cathode)} + E^0_{ox(anode)}$$

Examples of Redox Reactions

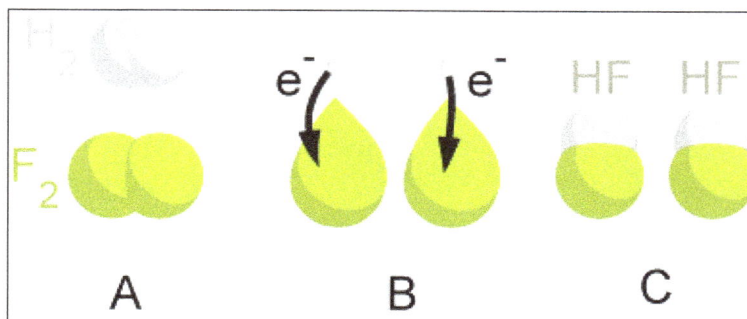

Illustration of a redox reaction.

A good example is the reaction between hydrogen and fluorine in which hydrogen is being oxidized and fluorine is being reduced:

$$H_2 + F_2 \rightarrow 2\,HF$$

We can write this overall reaction as two half-reactions:

1. The oxidation reaction:

$$H_2 \rightarrow 2H^+ + 2e^-$$

2. The reduction reaction:

$$F_2 + 2e^- \rightarrow 2F^-$$

Analyzing each half-reaction in isolation can often make the overall chemical process clearer. Because there is no net change in charge during a redox reaction, the number of electrons in excess in the oxidation reaction must equal the number consumed by the reduction reaction.

Elements, even in molecular form, always have an oxidation state of zero. In the first half-reaction, hydrogen is oxidized from an oxidation state of zero to an oxidation state of +1. In the second half-reaction, fluorine is reduced from an oxidation state of zero to an oxidation state of −1.

When adding the reactions together the electrons are canceled:

$$H_2 \rightarrow 2H^+ + 2e^-$$
$$\underline{F_2 + 2e^- \rightarrow 2F^-}$$
$$H_2 + F_2 \rightarrow 2H^+ + 2F^-$$

And the ions combine to form hydrogen fluoride:

$$2H^+ + 2F^- \rightarrow 2HF$$

The overall reaction is:

$$H_2 + F_2 \rightarrow 2HF$$

Metal Displacement

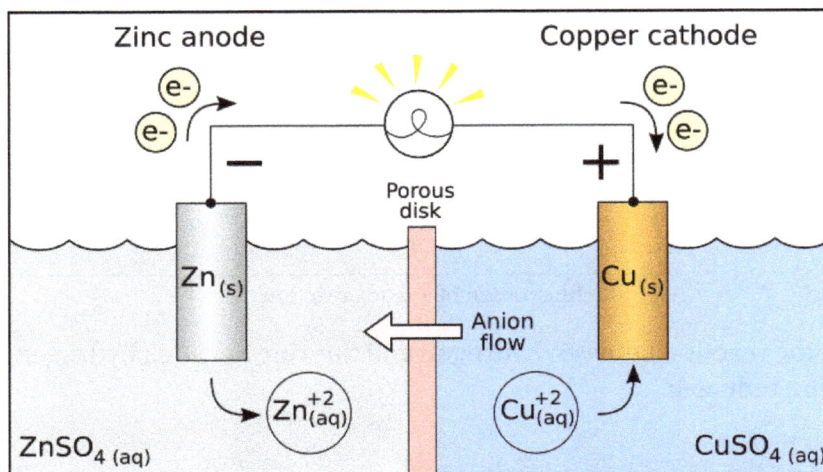

A redox reaction is the force behind an electrochemical cell like the Galvanic cell pictured. The battery is made out of a zinc electrode in a $ZnSO_4$ solution connected with a wire and a porous disk to a copper electrode in a $CuSO_4$ solution.

In this type of reaction, a metal atom in a compound (or in a solution) is replaced by an atom of another metal. For example, copper is deposited when zinc metal is placed in a copper(II) sulfate solution:

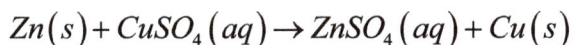

$$Zn(s) + CuSO_4(aq) \rightarrow ZnSO_4(aq) + Cu(s)$$

In the above reaction, zinc metal displaces the copper(II) ion from copper sulfate solution and thus liberates free copper metal.

The ionic equation for this reaction is:

$$Zn + Cu^{2+} \rightarrow Zn^{2+} + Cu$$

As two half-reactions, it is seen that the zinc is oxidized:

$$Zn \rightarrow Zn^{2+} + 2e^-$$

And the copper is reduced:

$$Cu^{2+} + 2e^- \rightarrow Cu$$

Other Examples

- The reduction of nitrate to nitrogen in the presence of an acid (denitrification):

$$2\,NO^{-}_{3} + 10\,e^{-} + 12\,H^{+} \rightarrow N_2 + 6\,H_2O$$

- The combustion of hydrocarbons, such as in an internal combustion engine, which produces water, carbon dioxide, some partially oxidized forms such as carbon monoxide, and heat energy. Complete oxidation of materials containing carbon produces carbon dioxide.

- In organic chemistry, the stepwise oxidation of a hydrocarbon by oxygen produces water and, successively, an alcohol, an aldehyde or a ketone, a carboxylic acid, and then a peroxide.

Corrosion and Rusting

Oxides, such as iron(III) oxide or rust, which consists of hydrated iron(III) oxides $Fe_2O_3 \cdot nH_2O$ and iron(III) oxide-hydroxide (FeO(OH), Fe(OH)3), form when oxygen combines with other elements.

Iron rusting in pyrite cubes.

- The term corrosion refers to the electrochemical oxidation of metals in reaction with an oxidant such as oxygen. Rusting, the formation of iron oxides, is a well-known example of electrochemical corrosion; it forms as a result of the oxidation of iron metal. Common rust often refers to iron(III) oxide, formed in the following chemical reaction:

$$4\,Fe + 3\,O_2 \rightarrow 2\,Fe_2O_3$$

- The oxidation of iron(II) to iron(III) by hydrogen peroxide in the presence of an acid:

$$Fe^{2+} \rightarrow Fe^{3+} + e^{-}$$
$$H_2O_2 + 2\,e^{-} \rightarrow 2\,OH^{-}$$

Overall equation:

$$2\,Fe^{2+} + H_2O_2 + 2\,H^{+} \rightarrow 2\,Fe^{3+} + 2\,H_2O$$

Redox Reactions in Industry

Cathodic protection is a technique used to control the corrosion of a metal surface by making it the cathode of an electrochemical cell. A simple method of protection connects protected metal to a more easily corroded "sacrificial anode" to act as the anode. The sacrificial metal instead of the protected metal, then, corrodes. A common application of cathodic protection is in galvanized steel, in which a sacrificial coating of zinc on steel parts protects them from rust.

Oxidation is used in a wide variety of industries such as in the production of cleaning products and oxidizing ammonia to produce nitric acid, which is used in most fertilizers.

Redox reactions are the foundation of electrochemical cells, which can generate electrical energy or support electrosynthesis. Metal ores often contain metals in oxidized states such as oxides or sulfides, from which the pure metals are extracted by smelting at high temperature in the presence of a reducing agent. The process of electroplating uses redox reactions to coat objects with a thin layer of a material, as in chrome-plated automotive parts, silver plating cutlery, galvanization and gold-plated jewelry.

Redox Reactions in Biology

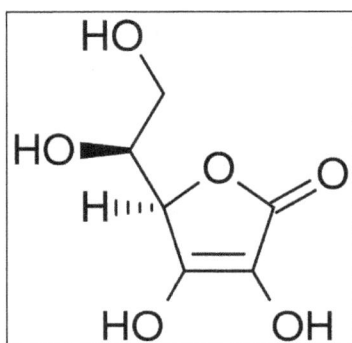

Ascorbic acid (reduced form of Vitamin C).   Dehydroascorbic acid (oxidized form of Vitamin C).

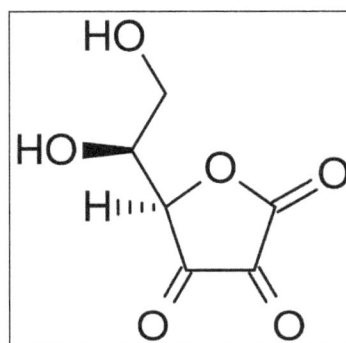

Many important biological processes involve redox reactions.

Cellular respiration, for instance, is the oxidation of glucose ($C_6H_{12}O_6$) to CO_2 and the reduction of oxygen to water. The summary equation for cell respiration is:

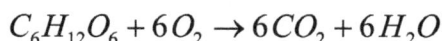

$$C_6H_{12}O_6 + 6O_2 \rightarrow 6CO_2 + 6H_2O$$

The process of cell respiration also depends heavily on the reduction of NAD$^+$ to NADH and the reverse reaction (the oxidation of NADH to NAD$^+$). Photosynthesis and cellular respiration are complementary, but photosynthesis is not the reverse of the redox reaction in cell respiration:

$$6CO_2 + 6H_2O + light\ energy \rightarrow C_6H_{12}O_6 + 6O_2$$

Biological energy is frequently stored and released by means of redox reactions. Photosynthesis involves the reduction of carbon dioxide into sugars and the oxidation of water into molecular oxygen. The reverse reaction, respiration, oxidizes sugars to produce carbon dioxide and water. As intermediate steps, the reduced carbon compounds are used to reduce nicotinamide adenine dinucleotide (NAD$^+$) to NADH, which then contributes to the creation of a proton gradient, which drives the synthesis of adenosine triphosphate (ATP) and is maintained by the reduction of oxygen. In animal cells, mitochondria perform similar functions.

Free radical reactions are redox reactions that occur as a part of homeostasis and killing microorganisms, where an electron detaches from a molecule and then reattaches almost instantaneously. Free radicals are a part of redox molecules and can become harmful to the human body if they do not reattach to the redox molecule or an antioxidant. Unsatisfied free radicals can spur the mutation of cells they encounter and are, thus, causes of cancer.

The term redox state is often used to describe the balance of GSH/GSSG, NAD$^+$/NADH and NADP$^+$/NADPH in a biological system such as a cell or organ. The redox state is reflected in the balance of several sets of metabolites (e.g., lactate and pyruvate, beta-hydroxybutyrate, and acetoacetate), whose interconversion is dependent on these ratios. An abnormal redox state can develop in a variety of deleterious situations, such as hypoxia, shock, and sepsis. Redox mechanism also control some cellular processes. Redox proteins and their genes must be co-located for redox regulation according to the CoRR hypothesis for the function of DNA in mitochondria and chloroplasts.

Redox Cycling

A wide variety of aromatic compounds are enzymatically reduced to form free radicals that contain one more electron than their parent compounds. In general, the electron donor is any of a wide variety of flavoenzymes and their coenzymes. Once formed, these anion free radicals reduce molecular oxygen to superoxide, and regenerate the unchanged parent compound. The net reaction is the oxidation of the flavoenzyme's coenzymes and the reduction of molecular oxygen to form superoxide. This catalytic behavior has been described as a futile cycle or redox cycling.

Redox Reactions in Geology

Mi Vida uranium mine, near Moab, Utah. The alternating red and white/green bands of sandstone correspond to oxidized and reduced conditions in groundwater redox chemistry.

In geology, redox is important to both the formation of minerals and the mobilization of minerals, and is also important in some depositional environments. In general, the redox state of most rocks can be seen in the color of the rock. The rock forms in oxidizing conditions, giving it a red color. It is then "bleached" to a green—or sometimes white—form when a reducing fluid passes through the rock. The reduced fluid can also carry uranium-bearing minerals. Famous examples of redox conditions affecting geological processes include uranium deposits and Moqui marbles.

Balancing Redox Reactions

Describing the overall electrochemical reaction for a redox process requires a balancing of the component half-reactions for oxidation and reduction. In general, for reactions in aqueous solution, this involves adding H^+, OH^-, H_2O and electrons to compensate for the oxidation changes.

Acidic Media

In acidic media, H⁺ ions and water are added to half-reactions to balance the overall reaction. For instance, when manganese(II) reacts with sodium bismuthate:

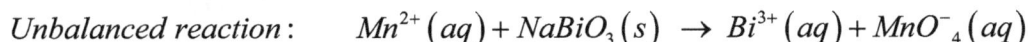

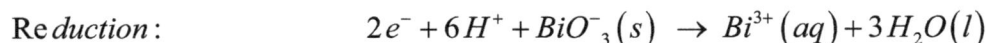

Unbalanced reaction: $Mn^{2+}(aq) + NaBiO_3(s) \rightarrow Bi^{3+}(aq) + MnO^-_4(aq)$

Oxidation: $4H_2O(l) + Mn^{2+}(aq) \rightarrow MnO^-_4(aq) + 8H^+(aq) + 5e^-$

Reduction: $2e^- + 6H^+ + BiO^-_3(s) \rightarrow Bi^{3+}(aq) + 3H_2O(l)$

The reaction is balanced by scaling the two half-cell reactions to involve the same number of electrons (multiplying the oxidation reaction by the number of electrons in the reduction step and vice versa):

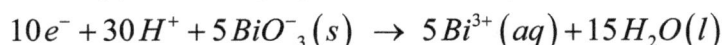

$$8H_2O(l) + 2Mn^{2+}(aq) \rightarrow 2MnO^-_4(aq) + 16H^+(aq) + 10e^-$$
$$10e^- + 30H^+ + 5BiO^-_3(s) \rightarrow 5Bi^{3+}(aq) + 15H_2O(l)$$

Adding these two reactions eliminates the electrons terms and yields the balanced reaction:

$$14H^+(aq) + 2Mn^{2+}(aq) + 5NaBiO_3(s) \rightarrow 7H_2O(l) + 2MnO^-_4(aq) + 5Bi^{3+}(aq) + 5Na^+(aq)$$

Basic Media

In basic media, OH⁻ ions and water are added to half reactions to balance the overall reaction. For example, in the reaction between potassium permanganate and sodium sulfite:

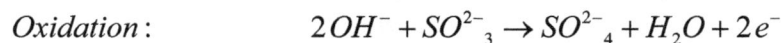

Unbalanced reaction: $KMnO_4 + Na_2SO_3 + H_2O \rightarrow MnO_2 + Na_2SO_4 + KOH$

Reduction: $3e^- + 2H_2O + MnO^-_4 \rightarrow MnO_2 + 4OH^-$

Oxidation: $2OH^- + SO^{2-}_3 \rightarrow SO^{2-}_4 + H_2O + 2e^-$

Balancing the number of electrons in the two half-cell reactions gives:

$$6e^- + 4H_2O + 2MnO^-_4 \rightarrow 2MnO_2 + 8OH^-$$
$$6OH^- + 3SO^{2-}_3 \rightarrow 3SO^{2-}_4 + 3H_2O + 6e^-$$

Adding these two half-cell reactions together gives the balanced equation:

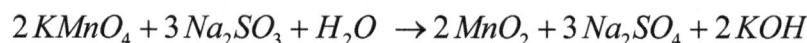

$$2KMnO_4 + 3Na_2SO_3 + H_2O \rightarrow 2MnO_2 + 3Na_2SO_4 + 2KOH$$

Mnemonics

The key terms involved in redox are often confusing. For example, a reagent that is oxidized loses electrons; however, that reagent is referred to as the reducing agent. Likewise, a reagent that is reduced gains electrons and is referred to as the oxidizing agent. These mnemonics are commonly used by students to help memorise the terminology:

- "OIL RIG": Oxidation is loss of electrons, reduction is gain of electrons.

- "LEO the lion says GER": Loss of electrons is oxidation, gain of electrons is reduction.

- "LEORA says GEROA": Loss of electrons is oxidation (reducing agent), gain of electrons is reduction (oxidizing agent).

- "RED CAT" and "AN OX", or "AnOx RedCat" ("an ox-red cat"): Reduction occurs at the cathode and the anode is for oxidation.

- "RED CAT gains what AN OX loses": Reduction at the cathode gains (electrons) what anode oxidation loses (electrons).

REDUCTION POTENTIAL

Redox potential (also known as oxidation/reduction potential, ORP, pe, ε, or E_h) is a measure of the tendency of a chemical species to acquire electrons from or lose electrons to an electrode and thereby be reduced or oxidised, respectively. Redox potential is measured in volts (V), or millivolts (mV). Each species has its own intrinsic redox potential; for example, the more positive the reduction potential (reduction potential is more often used due to general formalism in electrochemistry), the greater the species' affinity for electrons and tendency to be reduced. ORP can reflect the antimicrobial potential of the water.

Measurement and Interpretation

In aqueous solutions, redox potential is a measure of the tendency of the solution to either gain or lose electrons when it is subjected to change by introduction of a new species. A solution with a higher (more positive) reduction potential than the new species will have a tendency to gain electrons from the new species (i.e. to be reduced by oxidizing the new species) and a solution with a lower (more negative) reduction potential will have a tendency to lose electrons to the new species (i.e. to be oxidized by reducing the new species). Because the absolute potentials are next to impossible to accurately measure, reduction potentials are defined relative to a reference electrode. Reduction potentials of aqueous solutions are determined by measuring the potential difference between an inert sensing electrode in contact with the solution and a stable reference electrode connected to the solution by a salt bridge.

The sensing electrode acts as a platform for electron transfer to or from the reference half cell. It is typically platinum, although gold and graphite can be used as well. The reference half cell consists of a redox standard of known potential. The standard hydrogen electrode (SHE) is the reference from which all standard redox potentials are determined and has been assigned an arbitrary half cell potential of 0.0 mV. However, it is fragile and impractical for routine laboratory use. Therefore, other more stable reference electrodes such as silver chloride and saturated calomel (SCE) are commonly used because of their more reliable performance.

Although measurement of the redox potential in aqueous solutions is relatively straightforward, many factors limit its interpretation, such as effects of solution temperature and pH, irreversible reactions, slow electrode kinetics, non-equilibrium, presence of multiple redox couples, electrode poisoning, small exchange currents and inert redox couples. Consequently, practical measurements seldom correlate with calculated values. Nevertheless, reduction potential measurement has proven useful as an analytical tool in monitoring changes in a system rather than determining their absolute value (e.g. process control and titrations).

Similar to the concentration of hydrogen ion determines the acidity or pH of an aqueous solution, the tendency of electron transfer between a chemical species and an electrode determines the redox potential of an electrode couple. Like pH, redox potential represents how easily electrons are transferred to or from species in solution. Redox potential characterises the ability under the specific condition of a chemical species to lose or gain electrons instead of the amount of electrons available for oxidation or reduction.

In fact, it is possible to define pe, the negative logarithm of electron concentration (-log[e]) in a solution, which will be directly proportional to the redox potential. Sometimes pe is used as a unit of reduction potential instead of E_h, for example in environmental chemistry. If we normalize pe of hydrogen to zero, we will have the relation pe=16.9 E_h at room temperature. This point of view is useful for understanding redox potential, although the transfer of electrons, rather than the absolute concentration of free electrons in thermal equilibrium, is how one usually thinks of redox potential. Theoretically, however, the two approaches are equivalent.

Conversely, one could define a potential corresponding to pH as a potential difference between a solute and pH neutral water, separated by porous membrane (that is permeable to hydrogen ions). Such potential differences actually do occur from differences in acidity on biological membranes. This potential (where pH neutral water is set to 0 V) is analogous with redox potential (where standardized hydrogen solution is set to 0 V), but instead of hydrogen ions, electrons are transferred across in the redox case. Both pH and redox potentials are properties of solutions, not of elements or chemical compounds per se, and depend on concentrations, temperature etc.

Standard Reduction Potential

The standard reduction potential (E_0) is measured under standard conditions: 25 °C, a 1 activity for each ion participating in the reaction, a partial pressure of 1 bar for each gas that is part of the reaction, and metals in their pure state. The standard reduction potential is defined relative to a standard hydrogen electrode (SHE) reference electrode, which is arbitrarily given a potential of 0.00 V. However, because these can also be referred to as "redox potentials", the terms "reduction potentials" and "oxidation potentials" are preferred by the IUPAC. The two may be explicitly distinguished in symbols as E_0^r and E_0^o.

Half Cells

The relative reactivities of different half cells can be compared to predict the direction of electron flow. A higher E_0 means there is a greater tendency for reduction to occur, while a lower one means there is a greater tendency for oxidation to occur.

Any system or environment that accepts electrons from a normal hydrogen electrode is a half cell that is defined as having a positive redox potential; any system donating electrons to the hydrogen electrode is defined as having a negative redox potential. E_h is measured in millivolts (mV). A high positive E_h indicates an environment that favors oxidation reaction such as free oxygen. A low negative E_h indicates a strong reducing environment, such as free metals.

Sometimes when electrolysis is carried out in an aqueous solution, water, rather than the solute, is oxidized or reduced. For example, if an aqueous solution of NaCl is electrolyzed, water may be reduced at the cathode to produce $H_{2(g)}$ and OH^- ions, instead of Na^+ being reduced to $Na_{(s)}$, as occurs in the absence of water. It is the reduction potential of each species present that will determine which species will be oxidized or reduced.

Absolute reduction potentials can be determined if we find the actual potential between electrode and electrolyte for any one reaction. Surface polarization interferes with measurements, but various sources give an estimated potential for the standard hydrogen electrode of 4.4 V to 4.6 V (the electrolyte being positive).

Half-cell equations can be combined if one is reversed to an oxidation in a manner that cancels out the electrons to obtain an equation without electrons in it.

Nernst Equation

The E_h and pH of a solution are related. For a half cell equation, conventionally written as reduction (electrons on the left side):

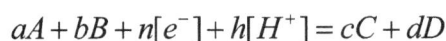

$$aA + bB + n[e^-] + h[H^+] = cC + dD$$

The half cell standard potential E_0 is given by:

$$E_0(volts) = -\frac{\Delta G^\ominus}{nF}$$

where $\Delta G^\ominus$ is the standard Gibbs free energy change, n is the number of electrons involved, and F is Faraday's constant. The Nernst equation relates pH and E_h:

$$E_h = E_0 + \frac{0.05916}{n}\log\left(\frac{\{A\}^a\{B\}^b}{\{C\}^c\{D\}^d}\right) - \frac{0.05916h}{n}\text{pH}$$

where curly brackets indicate activities and exponents are shown in the conventional manner. This equation is the equation of a straight line for E_h as a function of pH with a slope of volt (pH has no units.) This equation predicts lower E_h at higher pH values. This is observed for reduction of O_2 to OH^- and for reduction of H^+ to H_2. If H^+ were on the opposite side of the equation from H^+, the slope of the line would be reversed (higher E_h at higher pH). An example of that would be the formation of magnetite (Fe_3O_4) from $HFeO^-_{2\ (aq)}$:

$$3HFeO^-_2 + H^+ = Fe_3O_4 + 2\,H_2O + 2[[e^-]]$$

where $E_h = -1.1819 - 0.0885 \ \log([HFeO^-_2]^3) + 0.0296\text{pH}$. Note that the slope of the line is $-1/2$ the -0.05916 value above, since $h/n = -1/2$.

Biochemistry

Many enzymatic reactions are oxidation-reduction reactions in which one compound is oxidized and another compound is reduced. The ability of an organism to carry out oxidation-reduction reactions depends on the oxidation-reduction state of the environment, or its reduction potential (E_h).

Strictly aerobic microorganisms are generally active at positive E_h values, whereas strict anaerobes are generally active at negative E_h values. Redox affects the solubility of nutrients, especially metal ions.

There are organisms that can adjust their metabolism to their environment, such as facultative anaerobes. Facultative anaerobes can be active at positive E_h values, and at negative E_h values in the presence of oxygen bearing inorganic compounds, such as nitrates and sulfates.

Environmental Chemistry

In the field of environmental chemistry, the reduction potential is used to determine if oxidizing or reducing conditions are prevalent in water or soil, and to predict the states of different chemical species in the water, such as dissolved metals. pe values in water range from -12 to 25; the levels where the water itself becomes reduced or oxidized, respectively.

The reduction potentials in natural systems often lie comparatively near one of the boundaries of the stability region of water. Aerated surface water, rivers, lakes, oceans, rainwater and acid mine water, usually have oxidizing conditions (positive potentials). In places with limitations in air supply, such as submerged soils, swamps and marine sediments, reducing conditions (negative potentials) are the norm. Intermediate values are rare and usually a temporary condition found in systems moving to higher or lower pe values.

In environmental situations, it is common to have complex non-equilibrium conditions between a large number of species, meaning that it is often not possible to make accurate and precise measurements of the reduction potential. However, it is usually possible to obtain an approximate value and define the conditions as being in the oxidizing or reducing regime.

In the soil there are two main redox constituents: 1) anorganic redox systems (mainly ox/red compounds of Fe and Mn) and measurement in water extracts; 2) natural soil samples with all microbial and root components and measurement by direct method.

Water Quality

Oxidation reduction potential (ORP) can be used for water system monitoring with the benefit of a single-value measure of the disinfection potential, showing the activity of the disinfectant rather than the applied dose. For example, E. coli, Salmonella, Listeria and other pathogens have survival times of under 30 s when the ORP is above 665 mV, compared against >300 s when it is below 485 mV.

A study was conducted comparing traditional parts per million chlorination reading and ORP in Hennepin County, Minnesota. The results of this study argue for the inclusion of ORP above 650mV in local health codes.

Geology

E_h-pH (Pourbaix) diagrams are commonly used in mining and geology for assessment of the stability fields of minerals and dissolved species. Under the conditions where a mineral (solid) phase is predicted to be the most stable form of an element, these diagrams show that mineral. As the predicted results are all from thermodynamic (at equilibrium state) evaluations, these diagrams should be used with caution. Although the formation of a mineral or its dissolution may be predicted to occur under a set of conditions, the process may practically be negligible because its rate is too slow. Consequently, kinetic evaluations at the same time are necessary. Nevertheless, the equilibrium conditions can be used to evaluate the direction of spontaneous changes and the magnitude of the driving force behind them.

HALF-REACTION

A half reaction is either the oxidation or reduction reaction component of a redox reaction. A half reaction is obtained by considering the change in oxidation states of individual substances involved in the redox reaction.

Often, the concept of half-reactions is used to describe what occurs in an electrochemical cell, such as a Galvanic cell battery. Half-reactions can be written to describe both the metal undergoing oxidation (known as the anode) and the metal undergoing reduction (known as the cathode).

Half-reactions are often used as a method of balancing redox reactions. For oxidation-reduction reactions in acidic conditions, after balancing the atoms and oxidation numbers, one will need to add H^+ ions to balance the hydrogen ions in the half reaction. For oxidation-reduction reactions in basic conditions, after balancing the atoms and oxidation numbers, first treat it as an acidic solution and then add OH^- ions to balance the H^+ ions in the half reactions (which would give H_2O).

Example: Zn and Cu Galvanic cell.

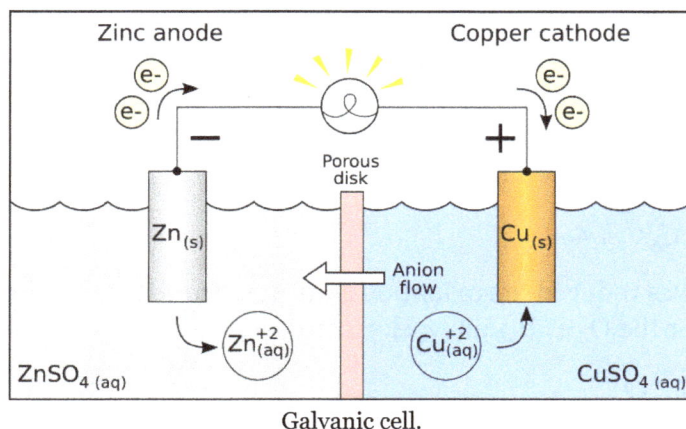

Galvanic cell.

Consider the Galvanic cell shown in the adjacent image: It is constructed with a piece of zinc (Zn) submerged in a solution of zinc sulfate ($ZnSO_4$) and a piece of copper (Cu) submerged in a solution of copper(II) sulfate ($CuSO_4$). The overall reaction is:

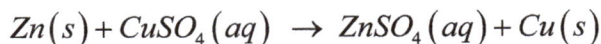

$$Zn(s) + CuSO_4(aq) \rightarrow ZnSO_4(aq) + Cu(s)$$

At the Zn anode, oxidation takes place (the metal loses electrons). This is represented in the following oxidation half-reaction (note that the electrons are on the products side):

$$Zn(s) \rightarrow Zn^{2+} + 2e^-$$

At the Cu cathode, reduction takes place (electrons are accepted). This is represented in the following reduction half-reaction (note that the electrons are on the reactants side):

$$Cu^{2+} + 2e^- \rightarrow Cu(s)$$

Example: oxidation of magnesium.

Photograph of a burning magnesium ribbon with
very short exposure to obtain oxidation detail.

Consider the example burning of magnesium ribbon (Mg). When magnesium burns, it combines with oxygen (O_2) from the air to form magnesium oxide (MgO) according to the following equation:

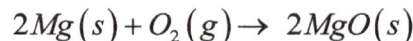

$$2Mg(s) + O_2(g) \rightarrow 2MgO(s)$$

Magnesium oxide is an ionic compound containing Mg^{2+} and O^{2-} ions whereas Mg(s) and O_2(g) are elements with no charges. The Mg(s) with zero charge gains a +2 charge going from the reactant side to product side, and the O_2(g) with zero charge gains a -2 charge. This is because when Mg(s) becomes Mg^{2+}, it loses 2 electrons. Since there are 2 Mg on left side, a total of 4 electrons are lost according to the following oxidation half reaction:

$$2Mg(s) \rightarrow 2Mg^{2+} + 4e^-$$

On the other hand, O_2 was reduced: its oxidation state goes from 0 to -2. Thus, a reduction half-reaction can be written for the O_2 as it gains 4 electrons:

$$O_2(g) + 4e^- \rightarrow 2O^{2-}$$

The overall reaction is the sum of both half-reactions:

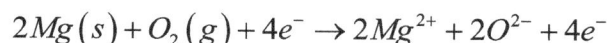

$$2Mg(s) + O_2(g) + 4e^- \rightarrow 2Mg^{2+} + 2O^{2-} + 4e^-$$

When chemical reaction, especially, redox reaction takes place, we do not see the electrons as they appear and disappear during the course of the reaction. What we see is the reactants (starting material) and end products. Due to this, electrons appearing on both sides of the equation are canceled. After canceling, the equation is re-written as:

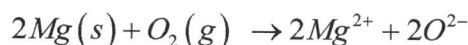

$$2Mg(s) + O_2(g) \rightarrow 2Mg^{2+} + 2O^{2-}$$

Two ions, positive (Mg^{2+}) and negative (O^{2-}) exist on product side and they combine immediately to form a compound magnesium oxide (MgO) due to their opposite charges (electrostatic attraction). In any given oxidation-reduction reaction, there are two half-reactions – oxidation half- reaction and reduction half-reaction. The sum of these two half-reactions is the oxidation- reduction reaction.

Half-reaction Balancing Method

Consider the reaction below:

$$Cl_2 + 2Fe^{2+} \rightarrow 2Cl^- + 2Fe^{3+}$$

The two elements involved, iron and chlorine, each change oxidation state; iron from +2 to +3, chlorine from 0 to −1. There are then effectively two half-reactions occurring. These changes can be represented in formulas by inserting appropriate electrons into each half-reaction:

$$Fe^{2+} \rightarrow Fe^{3+} + e^-$$
$$Cl_2 + 2e^- \rightarrow 2Cl^-$$

Given two half-reactions it is possible, with knowledge of appropriate electrode potentials, to arrive at the full (original) reaction the same way. The decomposition of a reaction into half-reactions is key to understanding a variety of chemical processes. For example, in the above reaction, it can be shown that this is a redox reaction in which Fe is oxidised, and Cl is reduced. Note the transfer of electrons from Fe to Cl. Decomposition is also a way to simplify the balancing of a chemical equation. A chemist can atom balance and charge balance one piece of an equation at a time.

For example:

$$Fe^{2+} \rightarrow Fe^{3+} + e^- \text{ becomes } 2Fe^{2+} \rightarrow 2Fe^{3+} + 2e^-$$

is added to $Cl_2 + 2e^- \rightarrow 2Cl^-$

and finally becomes $Cl_2 + 2Fe^{2+} \rightarrow 2Cl^- + 2Fe^{3+}$

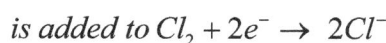

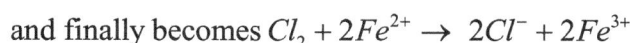

It is also possible and sometimes necessary to consider a half-reaction in either basic or acidic conditions, as there may be an acidic or basic electrolyte in the redox reaction. Due to this electrolyte it may be more difficult to satisfy the balance of both the atoms and charges. This is done by adding H_2O, OH^-, e^-, and or H+ to either side of the reaction until both atoms and charges are balanced.

Consider the half-reaction below:

$$PbO_2 \rightarrow PbO$$

OH^-, H_2O, and e^- can be used to balance the charges and atoms in basic conditions, as long as it is assumed that the reaction is in water.

$$2e^- + H_2O + PbO_2 \rightarrow PbO + 2OH^-$$

Again Consider the half-reaction below:

$$PbO_2 \rightarrow PbO$$

H^+, H_2O, and e^- can be used to balance the charges and atoms in acidic conditions, as long as it is assumed that the reaction is in water.

$$2e^- + 2H^+ + PbO_2 \rightarrow PbO + H_2O$$

Notice that both sides are both charge balanced and atom balanced.

Often there will be both H^+ and OH^- present in acidic and basic conditions but that the resulting reaction of the two ions will yield water H_2O:

$$H^+ + OH^- \rightarrow H_2O$$

COMBINATION REACTION

A combination reaction is a general category of chemical reactions. It may be defined as a chemical reaction in which two or more substances combine to form a single substance under suitable conditions.

Combination reactions are also known as synthesis, because in these reactions new substances are synthesized.

What are the Types of Combination Reaction?

Combination reactions are of three types.

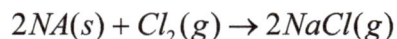

$$2NA(s) + Cl_2(g) \rightarrow 2NaCl(g)$$

Similarly, non-metals may react with highly active metals to form covalent compounds. Example, sulphur reacts with oxygen gas to form gaseous sulphur dioxide.

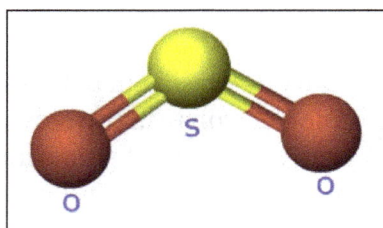

$$S(s) + O_2(g) \rightarrow SO_2(g)$$

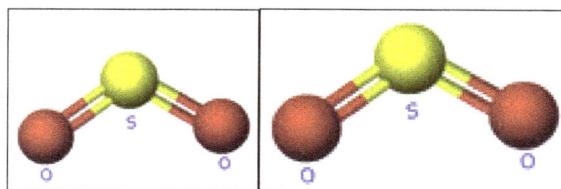

- Reaction between two or more elements: An example of this type of combination reaction is the reaction between a metal and a non- metal. Most metals react with non-metals to form ionic compounds. A good example of this would be:

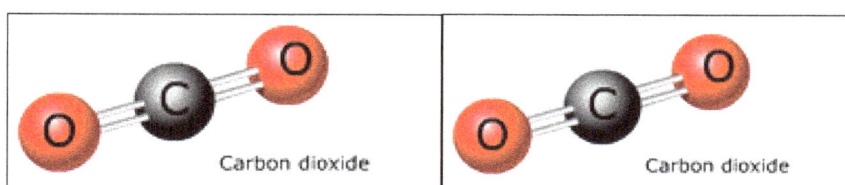

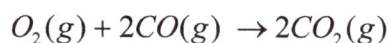

Carbon dioxide   Carbon dioxide

$$O_2(g) + 2CO(g) \rightarrow 2CO_2(g)$$

- Reaction between elements and compounds: An element and a compound react to form another compound. Example, carbon monoxide reacts with oxygen gas to form carbon dioxide.

- Reaction between two compounds: Two compounds react with each other to form a new compound. Example, calcium oxide (quick lime) reacts with carbon dioxide gas to form calcium carbonate (lime stone).

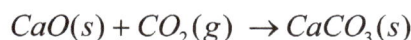

$$CaO(s) + CO_2(g) \rightarrow CaCO_3(s)$$

Why are most Combination Reactions are Exothermic in Nature?

Combination reactions involve the formation of new bonds and this process releases a large amount of energy in the form of heat.

- Formation of Calcium Hydroxide: Reaction between quick lime (Calcium oxide, CaO) and water is a combination reaction.

In this reaction, quick lime reacts with water to form slaked lime (calcium hydroxide,

Ca(OH)2). The reaction between quick lime and water is highly vigorous as well as exothermic.

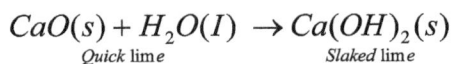

$$CaO(s) + H_2O(I) \rightarrow Ca(OH)_2(s)$$
$$\underset{Quick\ lime}{} \qquad\qquad \underset{Slaked\ lime}{}$$

- Combustion Reactions: Combustion of coal and combustion of hydrogen are examples of combination reactions.

Coal burns in air to form carbon dioxide gas.

$$C(s) + O_2(g) \rightarrow CO_2(g)$$

Hydrogen burns in the presence of oxygen to form water in the form of steam. Upon cooling, it becomes liquid.

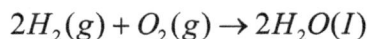

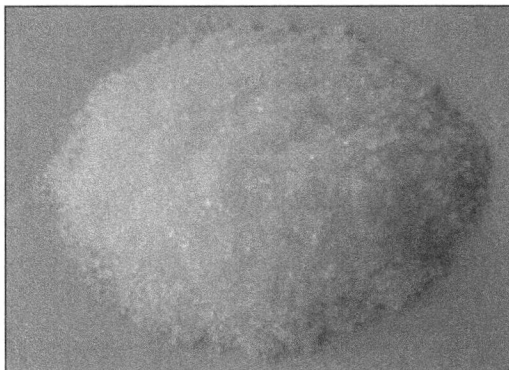

$$2H_2(g) + O_2(g) \rightarrow 2H_2O(I)$$

- Formation of Ammonium chloride: Ammonium chloride is formed by combining vapours of ammonia with hydrogen chloride gas. It is a white-coloured solid.

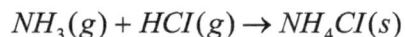

$$NH_3(g) + HCI(g) \rightarrow NH_4CI(s)$$

- Formation of Sulphuric acid: Formation of sulphuric acid from sulphur trioxide is also a combination reaction. Sulphur trioxide on hydration forms sulphuric acid. This reaction is highly exothermic in nature.

$$SO_3(g) + H_2O(I) \rightarrow H_2SO_4(I)$$

- Formation of ferrous sulphide: It is formed by heating fine pieces of iron with sulphur powder.

$$Fe(s) + S(s) \rightarrow FeS(s)$$

- Manufacture of Ammonia: The manufacture of ammonia gas from nitrogen and oxygen is also a combination reaction.

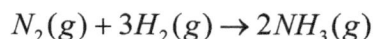

$$N_2(g) + 3H_2(g) \rightarrow 2NH_3(g)$$

- Combination of sodium oxide and water: Sodium oxide combines with water to form sodium hydroxide.

$$Na_2O(s) + H_2O(I) \rightarrow 2NaOH(s)$$

CORROSION REACTION

Corrosion reaction is an electrochemical reaction, wherein a material oxidises in the environment to form metal oxides.

Rust, the most familiar example of corrosion.

Volcanic gases have accelerated the extensive corrosion of this abandoned mining machinery, rendering it almost unrecognizable.

Corrosion is a natural process that converts a refined metal to a more chemically-stable form, such as its oxide, hydroxide, or sulfide. It is the gradual destruction of materials (usually metals) by chemical and electrochemical reaction with their environment. Corrosion engineering is the field dedicated to controlling and stopping corrosion.

Corrosion on exposed metal, including a bolt and nut.

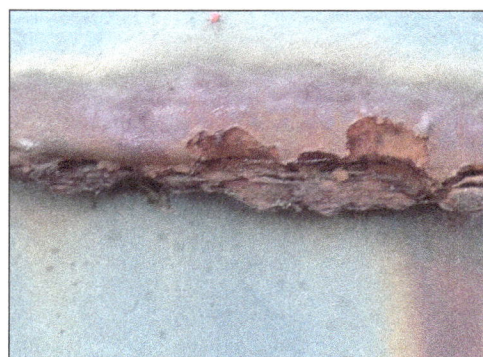

Side view Crow Hall Railway Bridge north of Preston Lancs corroding.

In the most common use of the word, this means electrochemical oxidation of metal in reaction with an oxidant such as oxygen or sulfates. Rusting, the formation of iron oxides, is a well-known example of electrochemical corrosion. This type of damage typically produces oxide(s) or salt(s) of the original metal and results in a distinctive orange colouration. Corrosion can also occur in materials other than metals, such as ceramics or polymers, although in this context, the term "degradation" is more common. Corrosion degrades the useful properties of materials and structures including strength, appearance and permeability to liquids and gases.

Many structural alloys corrode merely from exposure to moisture in air, but the process can be strongly affected by exposure to certain substances. Corrosion can be concentrated locally to form a pit or crack, or it can extend across a wide area more or less uniformly corroding the surface. Because corrosion is a diffusion-controlled process, it occurs on exposed surfaces. As a result, methods to reduce the activity of the exposed surface, such as passivation and chromate conversion, can increase a material's corrosion resistance. However, some corrosion mechanisms are less visible and less predictable.

Corrosion Removal

Often it is possible to chemically remove the products of corrosion. For example, phosphoric acid in the form of naval jelly is often applied to ferrous tools or surfaces to remove rust. Corrosion removal should not be confused with electropolishing, which removes some layers of the underlying metal to make a smooth surface. For example, phosphoric acid may also be used to electropolish copper but it does this by removing copper, not the products of copper corrosion.

Resistance to Corrosion

Some metals are more intrinsically resistant to corrosion than others (for some examples, see galvanic series). There are various ways of protecting metals from corrosion (oxidation) including painting, hot dip galvanizing, and combinations of these.

The materials most resistant to corrosion are those for which corrosion is thermodynamically unfavorable. Any corrosion products of gold or platinum tend to decompose spontaneously into pure metal, which is why these elements can be found in metallic form on Earth and have long been valued. More common "base" metals can only be protected by more temporary means.

Gold nuggets do not naturally corrode, even on a geological time scale.

Some metals have naturally slow reaction kinetics, even though their corrosion is thermodynamically favorable. These include such metals as zinc, magnesium, and cadmium. While corrosion of these metals is continuous and ongoing, it happens at an acceptably slow rate. An extreme example is graphite, which releases large amounts of energy upon oxidation, but has such slow kinetics that it is effectively immune to electrochemical corrosion under normal conditions.

Passivation

Passivation refers to the spontaneous formation of an ultrathin film of corrosion products, known as a passive film, on the metal's surface that act as a barrier to further oxidation. The chemical composition and microstructure of a passive film are different from the underlying metal. Typical passive film thickness on aluminium, stainless steels, and alloys is within 10 nanometers. The passive film is different from oxide layers that are formed upon heating and are in the micrometer thickness range – the passive film recovers if removed or damaged whereas the oxide layer does not. Passivation in natural environments such as air, water and soil at moderate pH is seen in such materials as aluminium, stainless steel, titanium, and silicon.

Passivation is primarily determined by metallurgical and environmental factors. The effect of pH is summarized using Pourbaix diagrams, but many other factors are influential. Some conditions that inhibit passivation include high pH for aluminium and zinc, low pH or the presence of chloride ions for stainless steel, high temperature for titanium (in which case the oxide dissolves into the metal, rather than the electrolyte) and fluoride ions for silicon. On the other hand, unusual conditions may result in passivation of materials that are normally unprotected, as the alkaline environment of concrete does for steel rebar. Exposure to a liquid metal such as mercury or hot solder can often circumvent passivation mechanisms.

Corrosion in Passivated Materials

Passivation is extremely useful in mitigating corrosion damage, however even a high-quality alloy will corrode if its ability to form a passivating film is hindered. Proper selection of the right grade of material for the specific environment is important for the long-lasting performance of this group of materials. If breakdown occurs in the passive film due to chemical or mechanical factors, the resulting major modes of corrosion may include pitting corrosion, crevice corrosion, and stress corrosion cracking.

Pitting Corrosion

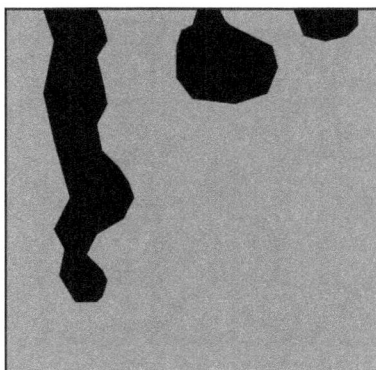

Diagram showing cross-section of pitting corrosion.

Certain conditions, such as low concentrations of oxygen or high concentrations of species such as chloride which compete as anions, can interfere with a given alloy's ability to re-form a passivating film. In the worst case, almost all of the surface will remain protected, but tiny local fluctuations will degrade the oxide film in a few critical points. Corrosion at these points will be greatly amplified, and can cause corrosion pits of several types, depending upon conditions. While the corrosion pits only nucleate under fairly extreme circumstances, they can continue to grow even when conditions return to normal, since the interior of a pit is naturally deprived of oxygen and locally the pH decreases to very low values and the corrosion rate increases due to an autocatalytic process. In extreme cases, the sharp tips of extremely long and narrow corrosion pits can cause stress concentration to the point that otherwise tough alloys can shatter; a thin film pierced by an invisibly small hole can hide a thumb sized pit from view. These problems are especially dangerous because they are difficult to detect before a part or structure fails. Pitting remains among the most common and damaging forms of corrosion in passivated alloys, but it can be prevented by control of the alloy's environment.

Pitting results when a small hole, or cavity, forms in the metal, usually as a result of de-passivation of a small area. This area becomes anodic, while part of the remaining metal becomes cathodic, producing a localized galvanic reaction. The deterioration of this small area penetrates the metal and can lead to failure. This form of corrosion is often difficult to detect due to the fact that it is usually relatively small and may be covered and hidden by corrosion-produced compounds.

Weld Decay and Knifeline Attack

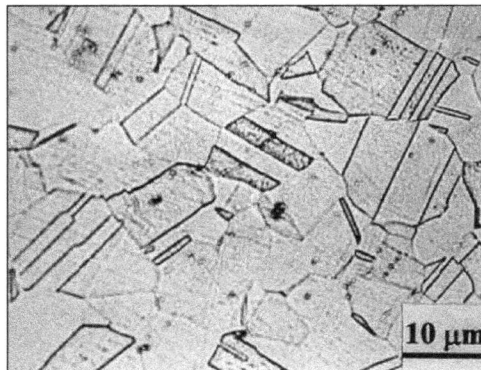

Normal microstructure of Type 304 stainless steel surface

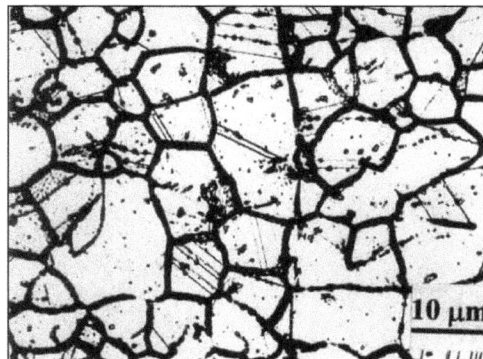

Sensitized metallic microstructure, showing
wider intergranular boundaries.

Stainless steel can pose special corrosion challenges, since its passivating behavior relies on the presence of a major alloying component (chromium, at least 11.5%). Because of the elevated temperatures of welding and heat treatment, chromium carbides can form in the grain boundaries of stainless alloys. This chemical reaction robs the material of chromium in the zone near the grain boundary, making those areas much less resistant to corrosion. This creates a galvanic couple with the well-protected alloy nearby, which leads to "weld decay" (corrosion of the grain boundaries in the heat affected zones) in highly corrosive environments. This process can seriously reduce the mechanical strength of welded joints over time.

A stainless steel is said to be "sensitized" if chromium carbides are formed in the microstructure. A typical microstructure of a normalized type 304 stainless steel shows no signs of sensitization, while a heavily sensitized steel shows the presence of grain boundary precipitates. The dark lines in the sensitized microstructure are networks of chromium carbides formed along the grain boundaries.

Special alloys, either with low carbon content or with added carbon "getters" such as titanium and niobium (in types 321 and 347, respectively), can prevent this effect, but the latter require special heat treatment after welding to prevent the similar phenomenon of "knifeline attack". As its name implies, corrosion is limited to a very narrow zone adjacent to the weld, often only a few micrometers across, making it even less noticeable.

Crevice Corrosion

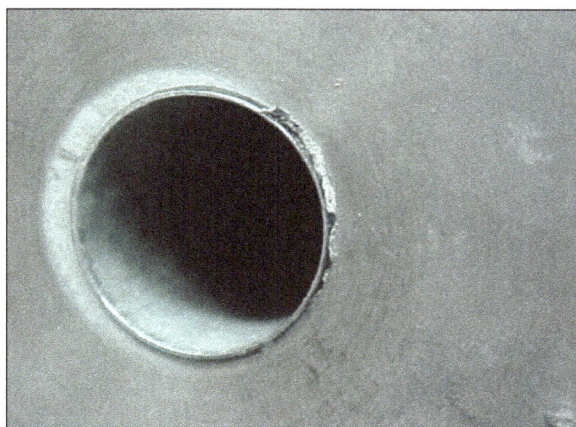

Corrosion in the crevice between the tube and tube sheet (both made of type 316 stainless steel) of a heat exchanger in a seawater desalination plant.

Crevice corrosion is a localized form of corrosion occurring in confined spaces (crevices), to which the access of the working fluid from the environment is limited. Formation of a differential aeration cell leads to corrosion inside the crevices. Examples of crevices are gaps and contact areas between parts, under gaskets or seals, inside cracks and seams, spaces filled with deposits and under sludge piles.

Crevice corrosion is influenced by the crevice type (metal-metal, metal-nonmetal), crevice geometry (size, surface finish), and metallurgical and environmental factors. The susceptibility to crevice corrosion can be evaluated with ASTM standard procedures. A critical crevice corrosion temperature is commonly used to rank a material's resistance to crevice corrosion.

Hydrogen Grooving

In the chemical industry, hydrogen grooving is the corrosion of piping by grooves created by the interaction of a corrosive agent, corroded pipe constituents, and hydrogen gas bubbles. For example, when sulfuric acid (H_2SO_4) flows through steel pipes, the iron in the steel reacts with the acid to form a passivation coating of iron sulfate ($FeSO_4$) and hydrogen gas (H_2). The iron sulfate coating will protect the steel from further reaction; however, if hydrogen bubbles contact this coating, it will be removed. Thus, a groove will be formed by a traveling bubble, exposing more steel to the acid: a vicious cycle. The grooving is exacerbated by the tendency of subsequent bubbles to follow the same path

High-temperature Corrosion

High-temperature corrosion is chemical deterioration of a material (typically a metal) as a result of heating. This non-galvanic form of corrosion can occur when a metal is subjected to a hot atmosphere containing oxygen, sulfur, or other compounds capable of oxidizing (or assisting the oxidation of) the material concerned. For example, materials used in aerospace, power generation and even in car engines have to resist sustained periods at high temperature in which they may be exposed to an atmosphere containing potentially highly corrosive products of combustion.

The products of high-temperature corrosion can potentially be turned to the advantage of the engineer. The formation of oxides on stainless steels, for example, can provide a protective layer preventing further atmospheric attack, allowing for a material to be used for sustained periods at both room and high temperatures in hostile conditions. Such high-temperature corrosion products, in the form of compacted oxide layer glazes, prevent or reduce wear during high-temperature sliding contact of metallic (or metallic and ceramic) surfaces. Thermal oxidation is also commonly used as a route towards the obtainment of controlled oxide nanostructures, including nanowires and thin films.

Microbial Corrosion

Microbial corrosion, or commonly known as microbiologically influenced corrosion (MIC), is a corrosion caused or promoted by microorganisms, usually chemoautotrophs. It can apply to both metallic and non-metallic materials, in the presence or absence of oxygen. Sulfate-reducing bacteria are active in the absence of oxygen (anaerobic); they produce hydrogen sulfide, causing sulfide stress cracking. In the presence of oxygen (aerobic), some bacteria may directly oxidize iron to iron oxides and hydroxides, other bacteria oxidize sulfur and produce sulfuric acid causing biogenic sulfide corrosion. Concentration cells can form in the deposits of corrosion products, leading to localized corrosion.

Accelerated low-water corrosion (ALWC) is a particularly aggressive form of MIC that affects steel piles in seawater near the low water tide mark. It is characterized by an orange sludge, which smells of hydrogen sulfide when treated with acid. Corrosion rates can be very high and design corrosion allowances can soon be exceeded leading to premature failure of the steel pile. Piles that have been coated and have cathodic protection installed at the time of construction are not susceptible to ALWC. For unprotected piles, sacrificial anodes can be installed locally to the affected areas to inhibit the corrosion or a complete retrofitted sacrificial anode system can be installed.

Affected areas can also be treated using cathodic protection, using either sacrificial anodes or applying current to an inert anode to produce a calcareous deposit, which will help shield the metal from further attack.

Metal Dusting

Metal dusting is a catastrophic form of corrosion that occurs when susceptible materials are exposed to environments with high carbon activities, such as synthesis gas and other high-CO environments. The corrosion manifests itself as a break-up of bulk metal to metal powder. The suspected mechanism is firstly the deposition of a graphite layer on the surface of the metal, usually from carbon monoxide (CO) in the vapor phase. This graphite layer is then thought to form metastable M_3C species (where M is the metal), which migrate away from the metal surface. However, in some regimes no M_3C species is observed indicating a direct transfer of metal atoms into the graphite layer.

Protection from Corrosion

The US military shrink wraps equipment such as helicopters to protect them from corrosion and thus save millions of dollars.

Various treatments are used to slow corrosion damage to metallic objects which are exposed to the weather, salt water, acids, or other hostile environments. Some unprotected metallic alloys are extremely vulnerable to corrosion, such as those used in neodymium magnets, which can spall or crumble into powder even in dry, temperature-stable indoor environments unless properly treated to discourage corrosion.

Surface Treatments

When surface treatments are used to retard corrosion, great care must be taken to ensure complete coverage, without gaps, cracks, or pinhole defects. Small defects can act as an "Achilles' heel", allowing corrosion to penetrate the interior and causing extensive damage even while the outer protective layer remains apparently intact for a period of time.

Applied Coatings

Plating, painting, and the application of enamel are the most common anti-corrosion treatments.

They work by providing a barrier of corrosion-resistant material between the damaging environment and the structural material. Aside from cosmetic and manufacturing issues, there may be tradeoffs in mechanical flexibility versus resistance to abrasion and high temperature. Platings usually fail only in small sections, but if the plating is more noble than the substrate (for example, chromium on steel), a galvanic couple will cause any exposed area to corrode much more rapidly than an unplated surface would. For this reason, it is often wise to plate with active metal such as zinc or cadmium. If the zinc coating is not thick enough the surface soon becomes unsightly with rusting obvious. The design life is directly related to the metal coating thickness.

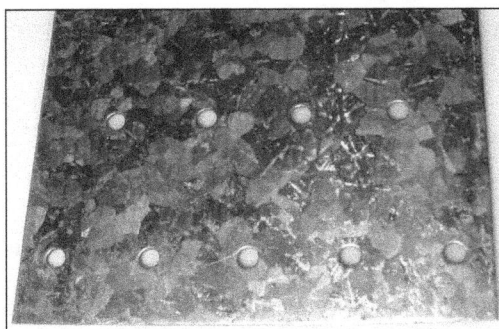

Galvanized surface.

Corroding Steel Electrification Gantry.

Painting either by roller or brush is more desirable for tight spaces; spray would be better for larger coating areas such as steel decks and waterfront applications. Flexible polyurethane coatings, like Durabak-M26 for example, can provide an anti-corrosive seal with a highly durable slip resistant membrane. Painted coatings are relatively easy to apply and have fast drying times although temperature and humidity may cause dry times to vary.

Reactive Coatings

If the environment is controlled (especially in recirculating systems), corrosion inhibitors can often be added to it. These chemicals form an electrically insulating or chemically impermeable coating on exposed metal surfaces, to suppress electrochemical reactions. Such methods make the system less sensitive to scratches or defects in the coating, since extra inhibitors can be made available wherever metal becomes exposed. Chemicals that inhibit corrosion include some of the salts in hard water (Roman water systems are famous for their mineral deposits), chromates, phosphates, polyaniline, other conducting polymers and a wide range of specially-designed chemicals that resemble surfactants (i.e. long-chain organic molecules with ionic end groups).

Anodization

This climbing descender is anodized with a yellow finish.

Aluminium alloys often undergo a surface treatment. Electrochemical conditions in the bath are carefully adjusted so that uniform pores, several nanometers wide, appear in the metal's oxide film. These pores allow the oxide to grow much thicker than passivating conditions would allow. At the end of the treatment, the pores are allowed to seal, forming a harder-than-usual surface layer. If this coating is scratched, normal passivation processes take over to protect the damaged area.

Anodizing is very resilient to weathering and corrosion, so it is commonly used for building facades and other areas where the surface will come into regular contact with the elements. While being resilient, it must be cleaned frequently. If left without cleaning, panel edge staining will naturally occur. Anodization is the process of converting an anode into cathode by bringing a more active anode in contact with it.

Biofilm Coatings

A new form of protection has been developed by applying certain species of bacterial films to the surface of metals in highly corrosive environments. This process increases the corrosion resistance substantially. Alternatively, antimicrobial-producing biofilms can be used to inhibit mild steel corrosion from sulfate-reducing bacteria.

Controlled Permeability Formwork

Controlled permeability formwork (CPF) is a method of preventing the corrosion of reinforcement by naturally enhancing the durability of the cover during concrete placement. CPF has been used in environments to combat the effects of carbonation, chlorides, frost and abrasion.

Cathodic Protection

Cathodic protection (CP) is a technique to control the corrosion of a metal surface by making that surface the cathode of an electrochemical cell. Cathodic protection systems are most commonly used to protect steel pipelines and tanks; steel pier piles, ships, and offshore oil platforms.

Sacrificial Anode Protection

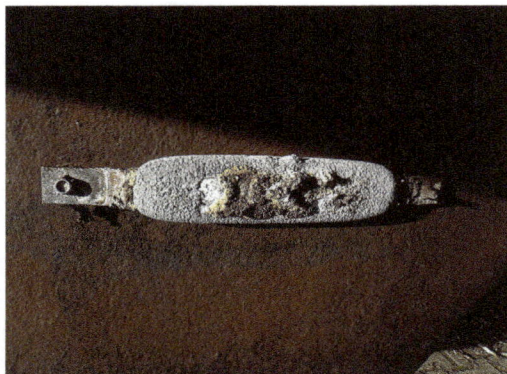

Sacrificial anode attached to the hull of a ship.

For effective CP, the potential of the steel surface is polarized (pushed) more negative until the metal surface has a uniform potential. With a uniform potential, the driving force for the corrosion reaction is halted. For galvanic CP systems, the anode material corrodes under the influence of the steel, and eventually it must be replaced. The polarization is caused by the current flow from the anode to the cathode, driven by the difference in electrode potential between the anode and the cathode. The most common sacrificial anode materials are aluminum, zinc, magnesium and related alloys. Aluminum has the highest capacity, and magnesium has the highest driving voltage and is thus used where resistance is higher. Zinc is general purpose and the basis for galvanizing.

Impressed Current Cathodic Protection

For larger structures, galvanic anodes cannot economically deliver enough current to provide complete protection. Impressed current cathodic protection (ICCP) systems use anodes connected to a DC power source (such as a cathodic protection rectifier). Anodes for ICCP systems are tubular and solid rod shapes of various specialized materials. These include high silicon cast iron, graphite, mixed metal oxide or platinum coated titanium or niobium coated rod and wires.

Anodic Protection

Anodic protection impresses anodic current on the structure to be protected (opposite to the cathodic protection). It is appropriate for metals that exhibit passivity (e.g. stainless steel) and suitably small passive current over a wide range of potentials. It is used in aggressive environments, such as solutions of sulfuric acid.

Rate of Corrosion

The formation of an oxide layer is described by the Deal Grove model, which is used to predict and control oxide layer formation in diverse situations. A simple test for measuring corrosion is the weight loss method. The method involves exposing a clean weighed piece of the metal or alloy to the corrosive environment for a specified time followed by cleaning to remove corrosion products and weighing the piece to determine the loss of weight. The rate of corrosion (R) is calculated as:

$$R = \frac{kW}{\rho At}$$

These neodymium magnets corroded extremely rapidly after only 5 months of outside exposure.

Where k is a constant, W is the weight loss of the metal in time t, A is the surface area of the metal exposed, and ρ is the density of the metal (in g/cm^3).

Other common expressions for the corrosion rate is penetration depth and change of mechanical properties.

Economic Impact

The collapsed Silver Bridge, as seen from the Ohio side.

In 2002, the US Federal Highway Administration released a study titled "Corrosion Costs and Preventive Strategies in the United States" on the direct costs associated with metallic corrosion in the US industry. In 1998, the total annual direct cost of corrosion in the U.S. was ca. $276 billion (ca. 3.2% of the US gross domestic product). Broken down into five specific industries, the economic losses are $22.6 billion in infrastructure; $17.6 billion in production and manufacturing; $29.7 billion in transportation; $20.1 billion in government; and $47.9 billion in utilities.

Rust is one of the most common causes of bridge accidents. As rust has a much higher volume than the originating mass of iron, its build-up can also cause failure by forcing apart adjacent parts. It was the cause of the collapse of the Mianus river bridge in 1983, when the bearings rusted internally and pushed one corner of the road slab off its support. Three drivers on the roadway at the time died as the slab fell into the river below. The following NTSB investigation showed that a drain in the road had been blocked for road re-surfacing, and had not been unblocked; as a result, runoff water penetrated the support hangers. Rust was also an important factor in the Silver Bridge disaster of 1967 in West Virginia, when a steel suspension bridge collapsed within a minute, killing 46 drivers and passengers on the bridge at the time.

Similarly, corrosion of concrete-covered steel and iron can cause the concrete to spall, creating severe structural problems. It is one of the most common failure modes of reinforced concrete bridges. Measuring instruments based on the half-cell potential can detect the potential corrosion spots before total failure of the concrete structure is reached.

Until 20–30 years ago, galvanized steel pipe was used extensively in the potable water systems for single and multi-family residents as well as commercial and public construction. Today, these systems have long ago consumed the protective zinc and are corroding internally resulting in poor water quality and pipe failures. The economic impact on homeowners, condo dwellers, and the public infrastructure is estimated at 22 billion dollars as the insurance industry braces for a wave of claims due to pipe failures.

Corrosion in Nonmetals

Most ceramic materials are almost entirely immune to corrosion. The strong chemical bonds that hold them together leave very little free chemical energy in the structure; they can be thought of as already corroded. When corrosion does occur, it is almost always a simple dissolution of the material or chemical reaction, rather than an electrochemical process. A common example of corrosion protection in ceramics is the lime added to soda-lime glass to reduce its solubility in water; though it is not nearly as soluble as pure sodium silicate, normal glass does form sub-microscopic flaws when exposed to moisture. Due to its brittleness, such flaws cause a dramatic reduction in the strength of a glass object during its first few hours at room temperature.

Corrosion of Polymers

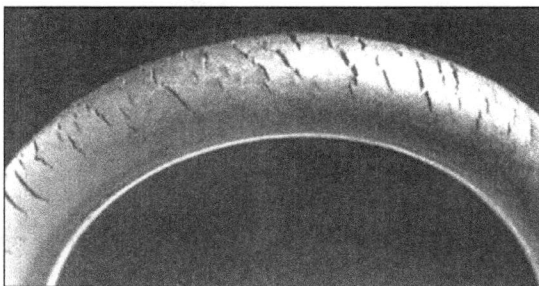

Ozone cracking in natural rubber tubing.

Polymer degradation involves several complex and often poorly understood physiochemical processes. These are strikingly different from the other processes discussed here, and so the term "corrosion" is only applied to them in a loose sense of the word. Because of their large molecular weight, very little entropy can be gained by mixing a given mass of polymer with another substance, making them generally quite difficult to dissolve. While dissolution is a problem in some polymer applications, it is relatively simple to design against.

A more common and related problem is "swelling", where small molecules infiltrate the structure, reducing strength and stiffness and causing a volume change. Conversely, many polymers (notably flexible vinyl) are intentionally swelled with plasticizers, which can be leached out of the structure, causing brittleness or other undesirable changes.

The most common form of degradation, however, is a decrease in polymer chain length. Mechanisms which break polymer chains are familiar to biologists because of their effect on DNA: ionizing radiation (most commonly ultraviolet light), free radicals, and oxidizers such as oxygen, ozone, and chlorine. Ozone cracking is a well-known problem affecting natural rubber for example. Plastic additives can slow these process very effectively, and can be as simple as a UV-absorbing pigment (e.g. titanium dioxide or carbon black). Plastic shopping bags often do not include these additives so that they break down more easily as ultrafine particles of litter.

Corrosion of Glass

Glass corrosion.

Glass is characterized by a high degree of corrosion-resistance. Because of its high water-resistance it is often used as primary packaging material in the pharma industry since most medicines are preserved in a watery solution. Besides its water-resistance, glass is also robust when exposed to certain chemically aggressive liquids or gases.

Glass disease is the corrosion of silicate glasses in aqueous solutions. It is governed by two mechanisms: diffusion-controlled leaching (ion exchange) and hydrolytic dissolution of the glass network. Both mechanisms strongly depend on the pH of contacting solution: the rate of ion exchange decreases with pH as $10^{-0.5pH}$ whereas the rate of hydrolytic dissolution increases with pH as $10^{0.5pH}$.

Mathematically, corrosion rates of glasses are characterized by normalized corrosion rates of elements NR_i (g/cm²·d) which are determined as the ratio of total amount of released species into the water M_i (g) to the water-contacting surface area S (cm²), time of contact t (days) and weight fraction content of the element in the glass f_i:

$$NR_i = \frac{M_i}{Sf_i t}.$$

The overall corrosion rate is a sum of contributions from both mechanisms (leaching + dissolution)

$NR_i = NRx_i + NRh$. Diffusion-controlled leaching (ion exchange) is characteristic of the initial phase of corrosion and involves replacement of alkali ions in the glass by a hydronium (H_3O^+) ion from the solution. It causes an ion-selective depletion of near surface layers of glasses and gives an inverse square root dependence of corrosion rate with exposure time. The diffusion-controlled normalized leaching rate of cations from glasses (g/cm²·d) is given by:

$$NRx_i = 2\rho\sqrt{\frac{D_i}{\pi t}},$$

where t is time, D_i is the i-th cation effective diffusion coefficient (cm²/d), which depends on pH of contacting water as $D_i = D_{i0}\cdot 10^{-pH}$, and ρ is the density of the glass (g/cm³).

Glass network dissolution is characteristic of the later phases of corrosion and causes a congruent release of ions into the water solution at a time-independent rate in dilute solutions (g/cm²·d):

$$NRh = \rho r_h,$$

where r_h is the stationary hydrolysis (dissolution) rate of the glass (cm/d). In closed systems the consumption of protons from the aqueous phase increases the pH and causes a fast transition to hydrolysis. However, a further saturation of solution with silica impedes hydrolysis and causes the glass to return to an ion-exchange, e.g. diffusion-controlled regime of corrosion.

In typical natural conditions normalized corrosion rates of silicate glasses are very low and are of the order of 10^{-7}–10^{-5} g/(cm²·d). The very high durability of silicate glasses in water makes them suitable for hazardous and nuclear waste immobilisation.

Glass Corrosion Tests

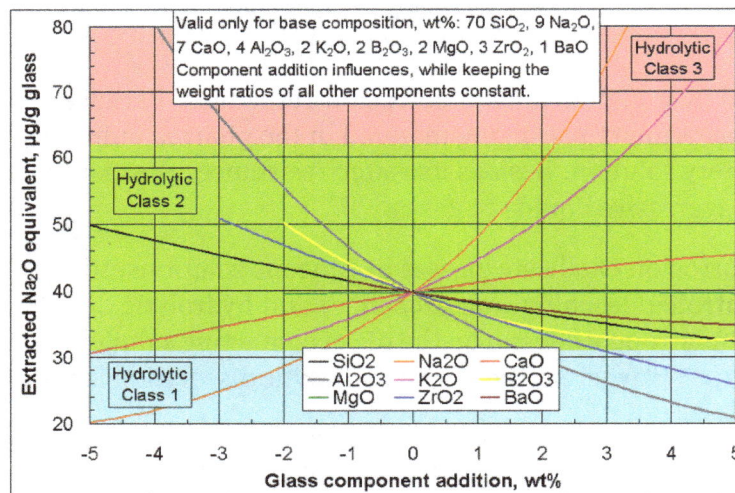

Effect of addition of a certain glass component on the chemical durability against water corrosion of a specific base glass (corrosion test ISO 719).

There exist numerous standardized procedures for measuring the corrosion (also called chemical durability) of glasses in neutral, basic, and acidic environments, under simulated environmental conditions, in simulated body fluid, at high temperature and pressure, and under other conditions.

The standard procedure ISO 719 describes a test of the extraction of water-soluble basic compounds under neutral conditions: 2 g of glass, particle size 300–500 μm, is kept for 60 min in 50 ml de-ionized water of grade 2 at 98 °C; 25 ml of the obtained solution is titrated against 0.01 mol/l HCl solution. The volume of HCl required for neutralization is classified according to the table below.

Amount of 0.01M HCl needed to neutralize extracted basic oxides, ml	Extracted Na_2O equivalent, μg	Hydrolytic class
< 0.1	< 31	1
0.1-0.2	31-62	2
0.2-0.85	62-264	3
0.85-2.0	264-620	4
2.0-3.5	620-1085	5
> 3.5	> 1085	> 5

The standardized test ISO 719 is not suitable for glasses with poor or not extractable alkaline components, but which are still attacked by water, e.g. quartz glass, B_2O_3 glass or P_2O_5 glass.

Usual glasses are differentiated into the following classes:

Hydrolytic class 1 (Type I):

This class, which is also called neutral glass, includes borosilicate glasses (e.g. Duran, Pyrex, Fiolax).

Glass of this class contains essential quantities of boron oxides, aluminium oxides and alkaline earth oxides. Through its composition neutral glass has a high resistance against temperature shocks and the highest hydrolytic resistance. Against acid and neutral solutions it shows high chemical resistance, because of its poor alkali content against alkaline solutions.

Hydrolytic class 2 (Type II):

This class usually contains sodium silicate glasses with a high hydrolytic resistance through surface finishing. Sodium silicate glass is a silicate glass, which contains alkali- and alkaline earth oxide and primarily sodium oxide and Calcium oxide.

Hydrolytic class 3 (Type III):

Glass of the 3rd hydrolytic class usually contains sodium silicate glasses and has a mean hydrolytic resistance, which is two times poorer than of type 1 glasses.

Acid class DIN 12116 and alkali class DIN 52322 (ISO 695) are to be distinguished from the hydrolytic class DIN 12111 (ISO 719).

Galvanic Corossion

Galvanic corrosion (also called bimetallic corrosion) is an electrochemical process in which one metal corrodes preferentially when it is in electrical contact with another, in the presence of an electrolyte. A similar galvanic reaction is exploited in primary cells to generate a useful electrical voltage to power portable devices.

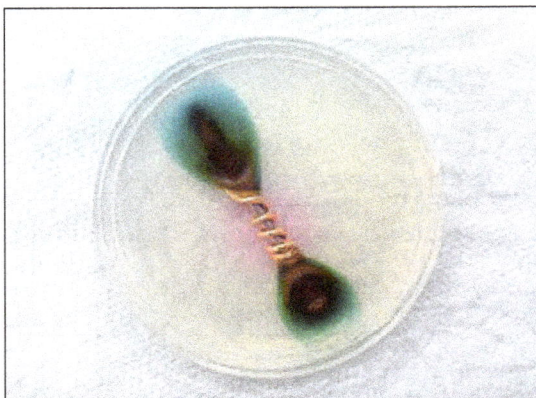

Corrosion of an iron nail wrapped in bright copper wire, showing cathodic protection of copper; a ferroxyl indicator solution shows colored chemical indications of two types of ions diffusing through a moist agar medium.

Dissimilar metals and alloys have different electrode potentials, and when two or more come into contact in an electrolyte, one metal acts as anode and the other as cathode. If the electrolyte contains only metal ions that are not easily reduced (such as Na^+, Ca^{2+}, K^+, Mg^{2+}, or Zn^{2+}), the cathode reaction is reduction of dissolved H^+ to H_2 or O_2 to OH^-. The electropotential difference between the reactions at the two electrodes is the driving force for an accelerated attack on the anode metal, which dissolves into the electrolyte. This leads to the metal at the anode corroding more quickly than it otherwise would and corrosion at the cathode being inhibited. The presence of an electrolyte and an electrical conducting path between the metals is essential for galvanic corrosion to occur. The electrolyte provides a means for ion migration whereby ions move to prevent charge build-up that would otherwise stop the reaction.

In some cases, this type of reaction is intentionally encouraged. For example, low-cost household batteries typically contain carbon-zinc cells. As part of a closed circuit (the electron pathway), the zinc within the cell will corrode preferentially (the ion pathway) as an essential part of the battery producing electricity. Another example is the cathodic protection of buried or submerged structures as well as hot water storage tanks. In this case, sacrificial anodes work as part of a galvanic couple, promoting corrosion of the anode, while protecting the cathode metal.

In other cases, such as mixed metals in piping (for example, copper, cast iron and other cast metals), galvanic corrosion will contribute to accelerated corrosion of parts of the system. Corrosion inhibitors such as sodium nitrite or sodium molybdate can be injected into these systems to reduce the galvanic potential. However, the application of these corrosion inhibitors must be monitored closely. If the application of corrosion inhibitors increases the conductivity of the water within the system, the galvanic corrosion potential can be greatly increased.

Acidity or alkalinity (pH) is also a major consideration with regard to closed loop bimetallic circulating systems. Should the pH and corrosion inhibition doses be incorrect, galvanic corrosion will be accelerated. In most HVAC systems, the use of sacrificial anodes and cathodes is not an option, as they would need to be applied within the plumbing of the system and, over time, would corrode and release particles that could cause potential mechanical damage to circulating pumps, heat exchangers, etc.

Examples of Corrosion

A common example of galvanic corrosion occurs in galvanized iron, a sheet of iron or steel covered with a zinc coating. Even when the protective zinc coating is broken, the underlying steel is not attacked. Instead, the zinc is corroded because it is less "noble"; only after it has been consumed can rusting of the base metal occur in earnest. By contrast, with a traditional tin can, the opposite of a protective effect occurs: because the tin is more noble than the underlying steel, when the tin coating is broken, the steel beneath is immediately attacked preferentially.

Statue of Liberty

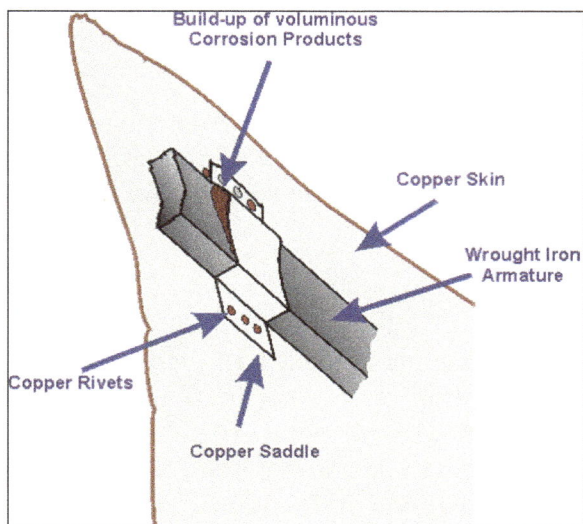

Galvanic corrosion in the Statue of Liberty.

Regular maintenance checks discovered that the Statue of Liberty suffered from galvanic corrosion.

A spectacular example of galvanic corrosion occurred in the Statue of Liberty when regular maintenance checks in the 1980s revealed that corrosion had taken place between the outer copper skin and the wrought iron support structure. Although the problem had been anticipated when the structure was built by Gustave Eiffel to Frédéric Bartholdi's design in the 1880s, the insulation layer of shellac between the two metals had failed over time and resulted in rusting of the iron supports. An extensive renovation was carried out requiring complete disassembly of the statue and replacement of the original insulation with PTFE. The structure was far from unsafe owing to the large number of unaffected connections, but it was regarded as a precautionary measure to preserve a national symbol of the United States.

Royal Navy and HMS Alarm

In the 17th century, Samuel Pepys (then serving as Admiralty Secretary) agreed to the removal of lead sheathing from English Royal Navy vessels to prevent the mysterious disintegration of their rudder-irons and bolt-heads, though he confessed himself baffled as to the reason the lead caused the corrosion.

The problem recurred when vessels were sheathed in copper to reduce marine weed accumulation and protect against shipworm. In an experiment, the Royal Navy in 1761 had tried fitting the hull of the frigate HMS Alarm with 12-ounce copper plating. Upon her return from a voyage to the West Indies, it was found that although the copper remained in fine condition and had indeed deterred shipworm, it had also become detached from the wooden hull in many places because the iron nails used during its installation "were found dissolved into a kind of rusty Paste". To the surprise of the inspection teams, however, some of the iron nails were virtually undamaged. Closer inspection revealed that water-resistant brown paper trapped under the nail head had inadvertently protected some of the nails: "Where this covering was perfect, the Iron was preserved from Injury". The copper sheathing had been delivered to the dockyard wrapped in the paper which was not always removed before the sheets were nailed to the hull. The conclusion therefore reported to the Admiralty in 1763 was that iron should not be allowed direct contact with copper in sea water.

US Navy Littoral Combat Ship Independence

Serious galvanic corrosion has been reported on the latest US Navy attack littoral combat vessel the USS Independence caused by steel water jet propulsion systems attached to an aluminium hull. Without electrical isolation between the steel and aluminium, the aluminium hull acts as an anode to the stainless steel, resulting in aggressive galvanic corrosion.

Corroding Lighting Fixtures

The unexpected fall in 2011 of a heavy light fixture from the ceiling of the Big Dig vehicular tunnel in Boston revealed that corrosion had weakened its support. Improper use of aluminum in contact with stainless steel had caused rapid corrosion in the presence of salt water. The electrochemical potential difference between stainless steel and aluminum is in the range of 0.5 to 1.0V, depending on the exact alloys involved, and can cause considerable corrosion within months under unfavorable conditions. Thousands of failing lights would have to be replaced, at an estimated cost of $54 million.

Lasagna Cell

A "lasagna cell" is accidentally produced when salty moist food such as lasagna is stored in a steel baking pan and is covered with aluminum foil. After a few hours the foil develops small holes where it touches the lasagna, and the food surface becomes covered with small spots composed of corroded aluminum. In this example, the salty food (lasagna) is the electrolyte, the aluminum foil is the anode, and the steel pan is the cathode. If the aluminum foil only touches the electrolyte in small areas, the galvanic corrosion is concentrated, and corrosion can occur fairly rapidly. If the aluminum foil was not used with a dissimilar metal container, the reaction was probably a chemical one. It is possible for heavy concentrations of salt, vinegar or some other acidic compound, or highly spiced foods to cause the foil to disintegrate. The product of either of these reactions is an aluminum salt. It does not harm the food, but any deposit may impart an undesired flavor and color.

Electrolytic Cleaning

The common technique of cleaning silverware by immersion of the silver or sterling silver (or even just silver plated objects) and a piece of aluminum (foil is preferred because of its much greater surface area compared to ingots, although if the foil has a "non-stick" face, this must be removed with steel wool first) in a hot electrolytic bath (usually composed of water and sodium bicarbonate, i.e., household baking soda) is an example of galvanic corrosion. Silver darkens and corrodes in the presence of airborne sulfur molecules, and the copper in sterling silver corrodes under a variety of conditions. These layers of corrosion can be largely removed through the electrochemical reduction of silver sulfide molecules: the presence of aluminum (which is less noble than either silver or copper) in the bath of sodium bicarbonate strips the sulfur atoms off the silver sulfide and transfers them onto and thereby corrodes the piece of aluminum foil (a much more reactive metal), leaving elemental silver behind. No silver is lost in the process.

Preventing Galvanic Corrosion

Aluminum anodes mounted on a steel-jacketed structure.

Electrical panel for a cathodic protection system.

There are several ways of reducing and preventing this form of corrosion:

- Electrically insulate the two metals from each other. If they are not in electrical contact, no galvanic coupling will occur. This can be achieved by using non-conductive materials between metals of different electropotential. Piping can be isolated with a spool of pipe made of plastic materials, or made of metal material internally coated or lined. It is important that the spool be a sufficient length to be effective. For reasons of safety, this should not be attempted where an electrical earthing system uses the pipework for its ground or has equipotential bonding.

- Metal boats connected to a shore line electrical power feed will normally have to have the hull connected to earth for safety reasons. However the end of that earth connection is likely to be a copper rod buried within the marina, resulting in a steel-copper "battery" of about 0.5 V. For such cases, the use of a galvanic isolator is essential, typically two semiconductor diodes in series, in parallel with two diodes conducting in the opposite direction (antiparallel). This prevents any current while the applied voltage is less than 1.4 V (i.e. 0.7 V per diode), but allows a full current in case of an electrical fault. There will still be a very minor leakage of current through the diodes, which may result in slightly faster corrosion than normal.

- Ensure there is no contact with an electrolyte. This can be done by using water-repellent compounds such as greases, or by coating the metals with an impermeable protective layer, such as a suitable paint, varnish, or plastic. If it is not possible to coat both, the coating should be applied to the more noble, the material with higher potential. This is advisable because if the coating is applied only on the more active material, in case of damage to the coating there will be a large cathode area and a very small anode area, and for the exposed anodic area the corrosion rate will be correspondingly high.

- Using antioxidant paste is beneficial for preventing corrosion between copper and aluminum electrical connections. The paste consists of a lower nobility metal than aluminum or copper.

- Choose metals that have similar electropotentials. The more closely matched the individual potentials, the smaller the potential difference and hence the smaller the galvanic current. Using the same metal for all construction is the easiest way of matching potentials.

- Electroplating or other plating can also help. This tends to use more noble metals that resist corrosion better. Chrome, nickel, silver and gold can all be used. Galvanizing with zinc protects the steel base metal by sacrificial anodic action.

- Cathodic protection uses one or more sacrificial anodes made of a metal which is more active than the protected metal. Alloys of metals commonly used for sacrificial anodes include zinc, magnesium, and aluminium. This approach is commonplace in water heaters and many buried or immersed metallic structures.

- Cathodic protection can also be applied by connecting a direct current (DC) electrical power supply to oppose the corrosive galvanic current.

Galvanic Series

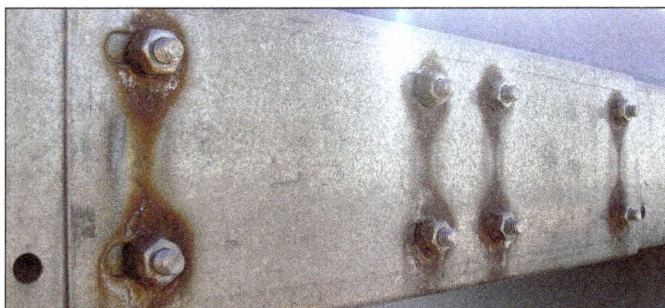

Galvanized mild steel cable ladder with corrosion around stainless steel bolts.

All metals can be classified into a galvanic series representing the electrical potential they develop in a given electrolyte against a standard reference electrode. The relative position of two metals on such a series gives a good indication of which metal is more likely to corrode more quickly. However, other factors such as water aeration and flow rate can influence the rate of the process markedly.

Anodic Index

The compatibility of two different metals may be predicted by consideration of their anodic index.

This parameter is a measure of the electrochemical voltage that will be developed between the metal and gold. To find the relative voltage of a pair of metals it is only required to subtract their anodic indices.

Sacrificial anode to protect a boat.

To reduce galvanic corrosion for metals stored in normal environments such as storage in warehouses or non-temperature and humidity controlled environments, there should not be more than 0.25 V difference in the anodic index of the two metals in contact. For controlled environments in which temperature and humidity are controlled, 0.50 V can be tolerated. For harsh environments such as outdoors, high humidity, and salty environments, there should be not more than 0.15 V difference in the anodic index. For example: gold and silver have a difference of 0.15V, therefore the two metals will not experience significant corrosion even in a harsh environment.

When design considerations require that dissimilar metals come in contact, the difference in anodic index is often managed by finishes and plating. The finishing and plating selected allow the dissimilar materials to be in contact, while protecting the more base materials from corrosion by the more noble. It will always be the metal with the most negative anodic index which will ultimately suffer from corrosion when galvanic incompatibility is in play. This is why sterling silver and stainless steel tableware should never be placed together in a dishwasher at the same time, as the steel items will likely experience corrosion by the end of the cycle (soap and water having served as the chemical electrolyte, and heat having accelerated the process).

Anaerobic Corrosion

Hydrogen corrosion is a form of metal corrosion occurring in the presence of anoxic water. Hydrogen corrosion involves a redox reaction that reduces hydrogen ions, forming molecular hydrogen.

Metals enter aqueous solution and are oxidized.

Oxidation reaction (pH independent):

$$Fe \rightarrow Fe^{2+} + 2e^{-}$$

Reduction reaction in acid solution:

$$2H^+ + 2e^- \rightarrow H_2$$

In an acidic solution, the water molecules are protonated and the hydronium ions (H_3O^+) are directly reduced into H_2.

Reduction reaction in neutral or slightly alkaline solution:

$$2H_2O + 2e^- \rightarrow H_2 + 2OH^-$$

In a neutral or slightly alkaline solution, the protons of water are reduced into molecular hydrogen giving rise to the production of hydroxide ions responsible of the precipitation of the slightly soluble ferrous hydroxide ($Fe(OH)_2$).

This finally leads to the global reaction of the anaerobic corrosion of iron in water:

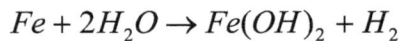

$$Fe + 2H_2O \rightarrow Fe(OH)_2 + H_2$$

Transformation of ferrous hydroxide into magnetite

Under anaerobic conditions, the ferrous hydroxide ($Fe(OH)_2$) can be oxidized by the protons of water to form magnetite and molecular hydrogen. This process is described by the Schikorr reaction:

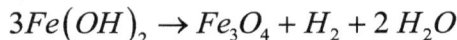

$$3Fe(OH)_2 \rightarrow Fe_3O_4 + H_2 + 2\ H_2O$$

ferrous hydroxide → magnetite + hydrogen + water

The well crystallized magnetite (Fe_3O_4) is thermodynamically more stable than the ferrous hydroxide ($Fe(OH)_2$).

This process also occurs during the anaerobic corrosion of iron and steel in oxygen-free groundwater and in reducing soils below the water table.

Corrosion Inhibitor

A corrosion inhibitor is a chemical compound that, when added to a liquid or gas, decreases the corrosion rate of a material, typically a metal or an alloy. The effectiveness of a corrosion inhibitor depends on fluid composition, quantity of water, and flow regime. A common mechanism for inhibiting corrosion involves formation of a coating, often a passivation layer, which prevents access of the corrosive substance to the metal. Permanent treatments such as chrome plating are not generally considered inhibitors, however. Instead corrosion inhibitors are additives to the fluids that surround the metal or related object.

Corrosion inhibitors are common in industry, and also found in over-the-counter products, typically in spray form in combination with a lubricant and sometimes a penetrating oil.

Types

The nature of the corrosive inhibitor depends on (i) the material being protected, which are most commonly metal objects, and (ii) on the corrosive agent to be neutralized. The corrosive agents are generally oxygen, hydrogen sulfide, and carbon dioxide. Oxygen is generally removed by reductive inhibitors such as amines and hydrazines:

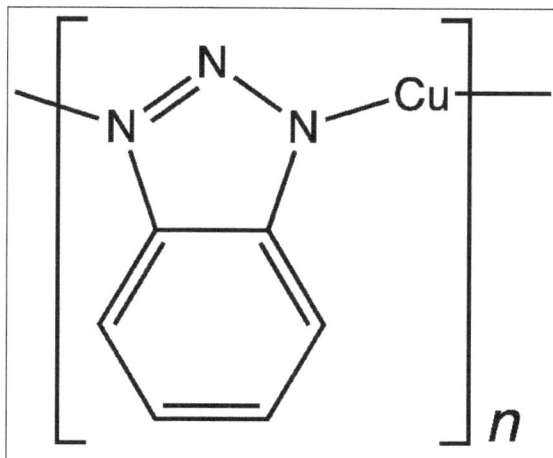

$$O_2 + N_2H_4 \rightarrow 2\,H_2O + N_2$$

Benzotriazole inhibits corrosion of copper by forming
an inert layer of this polymer on the metal's surface.

In this example, hydrazine converts oxygen, a common corrosive agent, to water, which is generally benign. Related inhibitors of oxygen corrosion are hexamine, phenylenediamine, and dimethylethanolamine, and their derivatives. Antioxidants such as sulfite and ascorbic acid are sometimes used. Some corrosion inhibitors form a passivating coating on the surface by chemisorption. Benzotriazole is one such species used to protect copper. For lubrication, zinc dithiophosphates are common - they deposit sulfide on surfaces.

The suitability of any given chemical for a task in hand depends on many factors, including their operating temperature.

Illustrative Applications

- Volatile amines are used in boilers to minimize the effects of acid. In some cases, the amines form a protective film on the steel surface and, at the same time, act as an anodic inhibitor. An inhibitor that acts both in a cathodic and anodic manner is termed a mixed inhibitor.

- Benzotriazole inhibits the corrosion and staining of copper surfaces.

- Corrosion inhibitors are often added to paints. A pigment with anticorrosive properties is zinc phosphate. Compounds derived from tannic acid or zinc salts of organonitrogens (e.g. Alcophor 827) can be used together with anticorrosive pigments. Other corrosion inhibitors are Anticor 70, Albaex, Ferrophos, and Molywhite MZAP.

- Antiseptics are used to counter microbial corrosion. Benzalkonium chloride is commonly used in oil field industry.

- In oil refineries, hydrogen sulfide can corrode steels so it is removed often using air and amines by conversion to polysulfides.

Fuels Industry

Corrosion inhibitors are commonly added to coolants, fuels, hydraulic fluids, boiler water, engine oil, and many other fluids used in industry. For fuels, various corrosion inhibitors can be used. Some components include zinc dithiophosphates.

- DCI-4A, widely used in commercial and military jet fuels, acts also as a lubricity additive. Can be also used for gasolines and other distillate fuels.

- DCI-6A, for motor gasoline and distillate fuels, and for U.S. military fuels (JP-4, JP-5, JP-8).

- DCI-11, for alcohols and gasolines containing oxygenates.

- DCI-28, for very low-pH alcohols and gasolines containing oxygenates.

- DCI-30, for gasoline and distillate fuels, excellent for pipeline transfers and storage, caustic-resistant.

- DMA-4 (solution of alkylaminophosphate in kerosene), for petroleum distillates.

Oil Additive

Oil additives are chemical compounds that improve the lubricant performance of base oil (or oil "base stock"). The manufacturer of many different oils can utilize the same base stock for each formulation and can choose different additives for each specific application. Additives comprise up to 5% by weight of some oils.

Nearly all commercial motor oils contain additives, whether the oils are synthetic or petroleum based. Essentially, only the American Petroleum Institute (API) Service SA motor oils have no additives, and they are therefore incapable of protecting modern engines. The choice of additives is determined by the application, e.g. the oil for a diesel engine with direct injection in a pickup truck (API Service CJ-4) has different additives than the oil used in a small gasoline-powered outboard motor on a boat (2-cycle engine oil).

Types of Additives

Oil additives are vital for the proper lubrication and prolonged use of motor oil in modern internal combustion engines. Without many of these, the oil would become contaminated, break down, leak out, or not properly protect engine parts at all operating temperatures. Just as important are additives for oils used inside gearboxes, automatic transmissions, and bearings. Some of the most important additives include those used for viscosity and lubricity, contaminant control, for the control of chemical breakdown, and for seal conditioning. Some additives permit lubricants to perform better under severe conditions, such as extreme pressures and temperatures and high levels of contamination.

Controlling Chemical Breakdown

- Detergent additives, dating back to the early 1930s, are used to clean and neutralize oil impurities which would normally cause deposits (oil sludge) on vital engine parts. Typical detergents are magnesium sulfonates.

- Corrosion or rust inhibiting additives retard the oxidation of metals inside an engine.

- Antioxidant additives retard the degradation of the oil stock by oxidation. Typical additives are organic amines and phenols.

- Metal deactivators create a film on metal surfaces to prevent the metal from causing the oil to be oxidized.

- Bases may be used to combat chemical decomposition of the base stock oil in the presence of acids. When oil is subjected to shear wear and oxidation by air and combustion gases, it will have a tendency to collect acids and increase its Total Acid Number (TAN). For example, the breakdown acids found in used gear oil may include carbocyclic acids, ketones, esters, and nitration and sulfation byproducts. However, organic and inorganic bases and detergents are included in most formulated oils, as discussed in the following paragraph, so some (but not all) of these contaminants will be neutralized. Gear oil degradation and longevity can be measured by its TAN.

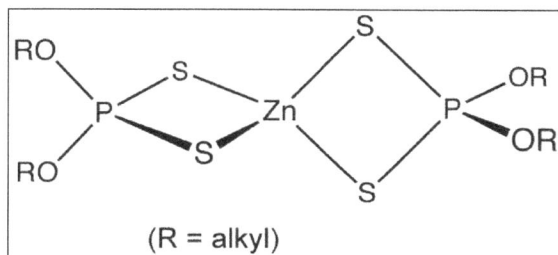

Chemical structure of a zinc dialkyldithiophosphate,
a typical antiwear agent found in many motor oils.

- Alkaline additives are used to neutralize the acids mentioned previously, and also help prevent the formation of sulfates in a working oil. A formulated oil will often have KOH (potassium hydroxide), a strong base, in small amounts, as it is an effective neutralizer used in refining petroleum. Additives that perform a similar function in a motor oil include magnesium and calcium sulphonates, salicylates, and phenates. These are the detergent additives mentioned previously. To measure the alkalinity potential of a formulated oil, it is tested to obtain the equivalent amount of KOH to arrive at the oil's Total Base Number (TBN) with units of mg of KOH per gram of oil. As the additive package degrades, TBN will decrease until the motor oil needs to be replaced. Further use of the oil will permit sludge, varnish, and metal corrosion. An important measurement of a motor oil's degradation and longevity is its TBN relative to a new oil.

For Viscosity

- Viscosity modifiers make an oil's viscosity higher at elevated temperatures, improving its viscosity index (VI). This combats the tendency of the oil to become thin at high temperature. The advantage of using less viscous oil with a VI improver is that it will have improved low temperature fluidity as well as being viscous enough to lubricate at operating temperature. Most multi-grade oils have viscosity modifiers. Some synthetic oils are engineered to meet multi-grade specifications without them. Viscosity modifiers are often plastic polymers. Virtually all oils require a specific range of viscosity as a working fluid, so viscosity is the primary factor that determines if an oil is acceptable for any particular application. As oils degrade from use, their viscosity will decrease, eventually requiring their replacement.

- Pour point depressants improve the oil's ability to flow at lower temperatures.

For Lubricity

- Friction modifiers or friction reducers, like molybdenum disulfide, are used for increasing fuel economy by reducing friction between moving parts. Friction modifiers alter the lubricity of the base oil. Whale oil was used historically.

- Extreme pressure agents bond to metal surfaces, keeping them from touching even at high pressure.

- Antiwear additives or wear inhibiting additives cause a film to surround metal parts, helping to keep them separated. Zinc dialkyldithiophosphate or zinc dithiophosphates are typically used.

- Nanoparticles that build diamond-like carbon coatings, which improve embeddability and can achieve Superlubricity. The technology is developed with Argonne National Lab and Pacific Northwest National Lab and foundation of TriboTEX product.

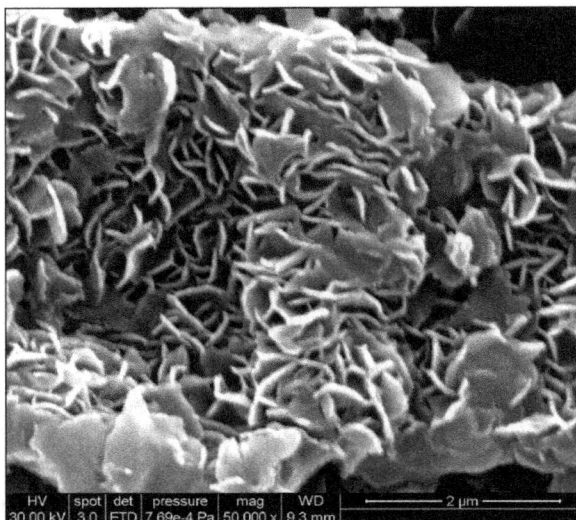

Nanoparticle flakes from the oil additive TriboTEX.
Image taken with electron microscope showing the nano scale.

- Inorganic Fullerene-like Tungsten Disulfide (IF-WS2) nanoparticles with a hollow sphere (Fullerene-like) morphology, provide extreme lubricity, anti-friction and high impact resistance (up to 35 GPa). The IF-WS2 particle was discovered by Professor Reshef Tenne at the Weizmann Institute of Science. Unlike standard lubricant additives that have platelet-like structures with moderate tribological properties, IF-WS2 particles (exclusively manufactured by ApNano - Nanotech Industrial Solutions) have tens of caged concentric layers, making these particles excel under extreme pressure or load. The IF-WS2 particles are available in dry powder form as well as a dispersion in oil, water, and solvent. These dispersions are used in the formulation of various lubricants, grease, metalworking fluids, coatings, paints, and polymers.

- Wear metals from friction are unintentional oil additives, but most large metal particles and impurities are removed in situ using either magnets or oil filters. Tribology is the science that studies how materials wear.

For Contaminant Control

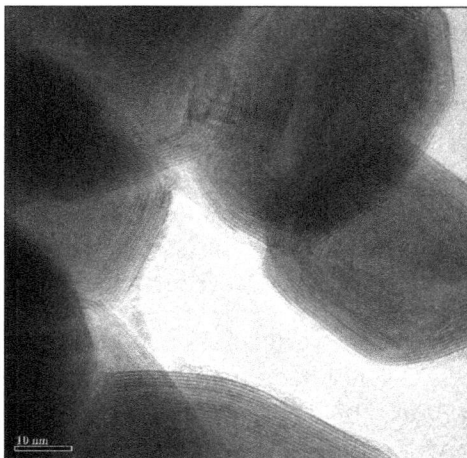

TEM image of group of scientific grade nanoparticles manufactured by Nanotech Industrial Solutions Corporation. Note the near spherical shape and presence of a hollow core:

- Dispersants keep contaminants (e.g. soot) suspended in the oil to prevent them from coagulating.

- Anti-foam agents (defoamants) inhibit the production of air bubbles and foam in the oil which can cause a loss of lubrication, pitting, and corrosion where entrained air and combustion gases contact metal surfaces.

- Antimisting agents prevent the atomization of the oil. Typical antimisting agents are silicones.

- Wax crystal modifiers are dewaxing aids that improve the ability of oil filters to separate wax from oil. This type of additive has applications in the refining and transport of oil, but not for lubricant formulation.

For other Reasons

- Seal conditioners cause gaskets and seals to swell to reduce oil leakage.

Additives in the Aftermarket and Controversy

Although motor oil is manufactured with numerous additives, aftermarket oil additives exist, too. A glaring inconsistency of mass-marketed aftermarket oil additives is that they often use additives which are foreign to motor oil. On the other hand, commercial additives are also sold that are designed for extended drain intervals (to replace depleted additives in used oil) or for formulating oils in situ (to make a custom motor oil from base stock). Commercial additives are identical to the additives found in off-the-shelf motor oil, while mass-marketed additives have some of each.

Some mass-market oil additives, notably the ones containing PTFE/Teflon (e.g. Slick 50) and chlorinated paraffins (e.g. Dura Lube), have caused a major backlash by consumers and the U.S. Federal Trade Commission which investigated many mass-marketed engine oil additives in the late 1990s.

Although there is no reason to say that all oil additives used in packaged engine oil are good and all aftermarket oil additives are bad, there has been a tendency in the aftermarket industry to make unfounded claims regarding the efficacy of their oil additives. These unsubstantiated claims have caused consumers to be lured into adding a bottle of chemicals to their engines which do not lower emissions, improve wear resistance, lower temperatures, improve efficiency, or extend engine life more than the (much cheaper) oil would have. Many consumers are convinced that aftermarket oil additives work, but many consumers are convinced that they do not work and are in fact detrimental to the engine. The topic is hotly debated on the Internet.

Although PTFE, a solid, was used in some aftermarket oil additives, users alleged that the PTFE clumped together, clogging filters. Certain people in the 1990s have reported that this was corroborated by NASA and U.S. universities. One thing to note, in defense of PTFE, is that if the particles are smaller than what was apparently used in the 1980s and 1990s, then PTFE can be an effective lubricant in suspension. The size of the particle and many other interrelated components of a lubricant make it difficult to make blanket statements about whether PTFE is useful or harmful. Although PTFE has been called "the slickest substance known to man", it would hardly do any good if it remains in the oil filter.

Lead(II,IV) Oxide

Lead(II,IV) oxide, also called red lead is the inorganic compound with the formula Pb_3O_4. A bright red or orange solid, it is used as pigment, in the manufacture of batteries, lead glass, and rustproof primer paints. It is an example of a mixed valence compound, being composed of both Pb(II) and Pb(IV).

Structure

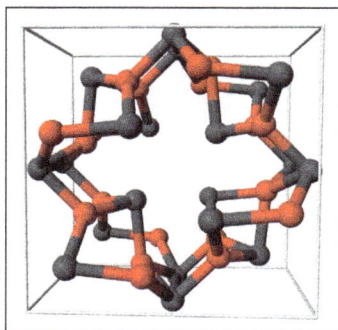

Unit cell of tetragonal Pb_3O_4 (Key: Pb O)

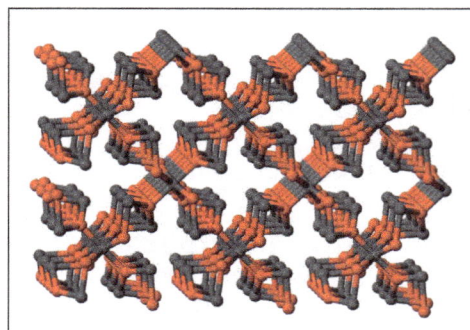

Part of tetragonal red lead's crystal structure

Lead(II,IV) oxide has a tetragonal crystal structure at room temperature, which then transforms to an orthorhombic (Pearson symbol oP28, Space group = Pbam, No 55) form at temperature 170 K (−103 °C). This phase transition only changes the symmetry of the crystal and slightly modifies the interatomic distances and angles.

Preparation

Lead(II,IV) oxide is prepared by calcination of lead(II) oxide (PbO; also called litharge) in air at about 450-480 °C:

$$6\,PbO + O_2 \rightarrow 2\,Pb_3O_4$$

The resulting material is contaminated with PbO. If a pure compound is desired, PbO can be removed by a potassium hydroxide solution:

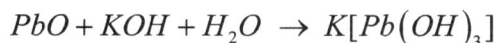

$$PbO + KOH + H_2O \rightarrow K[Pb(OH)_3]$$

Another method of preparation relies on annealing of lead(II) carbonate (cerussite) in air:

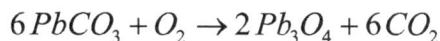

$$6\,PbCO_3 + O_2 \rightarrow 2\,Pb_3O_4 + 6\,CO_2$$

Yet another method is oxidative annealing of white lead:

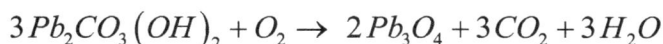

$$3\,Pb_2CO_3(OH)_2 + O_2 \rightarrow 2\,Pb_3O_4 + 3\,CO_2 + 3\,H_2O$$

In solution, lead(II,IV) oxide can be prepared by reaction of potassium plumbate with lead(II) acetate, yielding yellow insoluble lead(II,IV) oxide monohydrate, $Pb_3O_4 \cdot H_2O$, which can be turned into the anhydrous form by gentle heating:

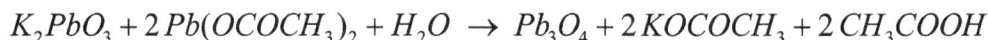

$$K_2PbO_3 + 2\,Pb(OCOCH_3)_2 + H_2O \rightarrow Pb_3O_4 + 2\,KOCOCH_3 + 2\,CH_3COOH$$

Natural minium is uncommon, forming only in extreme oxidizing conditions of lead ore bodies. The best known natural specimens come from Broken Hill, New South Wales, Australia, where they formed as the result of a mine fire.

Reactions

Red lead is virtually insoluble in water and in ethanol. However, it is soluble in hydrochloric acid present in the stomach, and is therefore toxic when ingested. It also dissolves in glacial acetic acid and a diluted mixture of nitric acid and hydrogen peroxide.

When heated to 500 °C, it decomposes to lead(II) oxide and oxygen. At 580 °C, the reaction is complete.

$$2\,Pb_3O_4 \rightarrow 6\,PbO + O_2$$

Nitric acid dissolves the lead(II) oxide component, leaving behind the insoluble lead(IV) oxide:

$$Pb_3O_4 + 4\,HNO_3 \rightarrow PbO_2 + 2\,Pb(NO_3)_2 + 2\,H_2O$$

With iron oxides and with elemental iron, lead(II,IV) oxide forms insoluble iron(II) and iron(III) plumbates, which is the basis of the anti-corrosive properties of lead-based paints applied to iron objects.

Use

Lead tetroxide is most often used as a pigment for primer paints for iron objects. Due to its toxicity, its use is being limited. In the past, it was used in combination with linseed oil as a thick, long-lasting anti-corrosive paint. The combination of minium and linen fibres was also used for plumbing, now replaced with PTFE tape. Currently it is mostly used for manufacture of glass, especially lead crystal glass. It finds limited use in some amateur pyrotechnics as a delay charge and was used in the past in the manufacture of dragon's egg pyrotechnic stars.

Red lead is used as a curing agent in some polychloroprene rubber compounds. It is used in place of magnesium oxide to provide better water resistance properties.

Red lead was used for engineer's scraping, before being supplanted by engineer's blue.

It is also used as an adultering agent in turmeric powder.

Physiological Effects

When inhaled, lead(II,IV) oxide irritates lungs. In case of high dose, the victim experiences a metallic taste, chest pain, and abdominal pain. When ingested, it is dissolved in the gastric acid and absorbed, leading to lead poisoning. High concentrations can be absorbed through skin as well, and it is important to follow safety precautions when working with lead-based paint.

Long-term contact with lead(II,IV) oxide may lead to accumulation of lead compounds in organisms, with development of symptoms of acute lead poisoning. Chronic poisoning displays as agitation, irritability, vision disorders, hypertension, and a grayish facial hue.

Lead(II,IV) oxide was shown to be carcinogenic for laboratory animals. Its carcinogenicity for humans was not proven.

Volatile Corrosion Inhibitor

Volatile corrosion inhibitors (VCI) are a type of corrosion inhibitor that are used to protect ferrous materials and non ferrous metals against corrosion or oxidation where it is impractical to apply surface treatments. They slowly release chemical compounds within a sealed airspace that actively prevents surface corrosion. A typical application is to protect stored tools or parts inside bags, boxes or cupboards, one advantage of VCIs being that if the container is opened and reclosed, levels of inhibitor will recover.

Also known as Vapor Phase Corrosion Inhibitors (VpCIs) or metal air inhibitors, vapor phase inhibitors (VPI) began in the 1940s when Shell Petroleum developed the very first of the traditional VCI's using a chemical compound called DICHAN or Dicyclohexylammonium Nitrite. This was used by the US military to prevent various metal components from corrosion and used it in multiple formats such as VCI paper, VCI powder, VCI solutions etc. Due to the dangerous nature of the chemistry DICHAN is now a mostly banned substance and there was a distinct break in the development of VCI's into two major groups of nitrite based VCI's and Amine based VCI's. The inclusion of either nitrite, secondary and tertiary amines is now frowned upon in the packaging world as combinations of these types of chemicals can cause Nitrosamines which are cancer forming agents.

VCI chemicals dissociate into anions and cations with moisture as an electrolyte. These anions and cations associate with anodic and cathodic reactive areas respectively on the metal surface forming a mono-ionic protective layer, making it inert for any further reaction such as oxidation/ rusting. Any loss of ions is replenished through further condensation of vapor ensuring continuous protection.

The report Nitrosation of Volatile Amines at the Workplace. The MAK Collection for Occupational Health and Safety. Nitrosation also can occur to primary amines: "N-Nitrosamines are formed in the nitrosation reaction from primary, secondary, and tertiary amines. The end product of the nitrosation of aromatic primary amines is a diazonium salt; with primary alkyl amines a mixture of products is obtained which can also contain small amounts of dialkyl nitrosamines. In all these cases the first and rate-limiting step is an electrophilic attack by the nitrosating agent (NO·Y) on the free electron pair of the amine nitrogen. The amine is preferentially attacked in its unprotonated, basic form.

Amines can be converted to nitrosamines by reaction with nitrosation agents. Such nitrosation reactions can take place even under conditions which are far from optimal and, with few exceptions, the nitrosamines produced have proved in animal studies to be very potent carcinogens. Of the about 300 different N-nitroso compounds which have been tested to date, 90 % have been shown to have carcinogenic activity in about 40 species of animals; no animal species is resistant. The doses required for tumour induction are extremely small".

In July 29, 2004, Lynn Kenison, Senior Chemist for the USDOL OSHA Salt Lake Technical Center, wrote a report on his findings about VCIs and the need to include Material Safety Data Sheets (MSDSs). His finding included the following:

1. VCIs are Hazardous Chemicals and do fall under the OSHA Hazard Communication Standard – 29 CFR 1910.1200. This means manufacturers must comply with the comprehensive hazard communication program explained above.

2. VCIs may present a Physical Hazard, a Health Hazard or both a Physical and Health Hazard. (a) Physical Hazard – means a chemical for which there is scientifically valid evidence that it is a combustible liquid, a compressed gas, explosive, flammable, an organic peroxide, an oxidizer, pyrophoric, unstable (reactive) or water-reactive. (b) Health Hazard – means a chemical for which there is statistically significant evidence based on at least one study conducted in accordance with established scientific principles that acute or chronic health effects may occur in exposed employees. The term "health hazard" includes chemicals which are carcinogens, toxic or highly toxic agents, reproductive toxins, irritants, corrosives, sensitizers, hepatotoxins, nephrotoxins, neurotoxins, agents which act on the hematopoietic system and agents which damage the lungs, skin, eyes, or mucous membranes.

3. VCIs do not meet the definition of "articles," but they do meet the definition of "Hazardous Chemicals".

4. Trade Secrets – The chemical manufacturer, importer, or employer may withhold the specific chemical identity, including the chemical name and other specific identification of a hazardous chemical, from the material safety data sheet, provided that: (a) The claim that the information withheld is a trade secret can be supported. (b) Information contained in the material data sheet concerning the properties and effects of the hazardous chemical is disclosed. (c) The material safety data sheet indicates that the specific chemical identity is being withheld as a trade secret. (d) The specific chemical identity is made available to health professionals, employees, and designated representatives in accordance with the applicable provisions of this paragraph. (e) The name(s) of all other chemical in the product is/are revealed along with their hazards.

5. Claiming ignorance ("I don't know what is in this material and I don't want to know!") to avoid accountability, does not eliminate responsibility. Manufacturers are still responsible for the correct identification and warnings listed for each VCI.

6. ISO 9001 and ISO 14001: Apply to the processes that an organization employs to realize its products or services. In other words, the way it accomplishes its work and meets the customers' requirements. ISO 9001 applies to the processes that influence product or service quality, and ISO 14001 applies to the processes that influence the organization's environmental performance. ISO 9001 and ISO 14001 have nothing to do with "Product Quality" or "Product Guarantee".

Product Uses

VCI chemicals are often added to paper and plastic substrates as a medium to deliver the protective chemical compounds for use in automotive packaging, steel packaging, metal packaging, military and hobby markets.

Product Limitations

According to the US Military Packaging Manual:

Restrictions and limitations in the use of VCI. VCI materials will not protect all metals from corrosion; in fact, they appear to increase the rate of corrosion in certain metals. VCI materials must not be used to protect any assemblies containing optical systems or precision moving parts which have been coated with a preservative or lubricant, unless otherwise specified. Items protected with bonded films, such as molybdenum (a dry lubricant), are not included in this category. VCI materials are affected by heat and light. They lose their effectiveness as the temperature increases and they decompose if exposed to direct sunlight for extended periods. They also decompose in the presence of acids or strong alkalies. Precautions must be taken when VCI is used with items, assemblies, and subassemblies containing zinc plate, cadmium, zinc. base alloys, magnesium base alloys, lead base alloys, and alloys of other metals including solders and brazing alloys. If such items contain more than 30 percent of zinc or 9 percent of lead, they must not be preserved with VCI. In all cases direct contact of VCI with non-ferrous metals except aluminum and aluminum-base alloys should be avoided unless specific permission has been granted. Care should also be taken with assemblies containing plastics, painted parts, or components of natural or synthetic rubber. Assemblies containing parts made of these materials should not be packed with VCI until proof is established that they have passed the compatibility test required by Specification.

Black Oxide

Black oxide or blackening is a conversion coating for ferrous materials, stainless steel, copper and copper based alloys, zinc, powdered metals, and silver solder. It is used to add mild corrosion resistance, for appearance and to minimize light reflection. To achieve maximal corrosion resistance the black oxide must be impregnated with oil or wax. One of its advantages over other coatings is its minimal buildup.

Ferrous Materials

A standard black oxide is magnetite (Fe_3O_4), which is more mechanically stable on the surface and provides better corrosion protection than red oxide (rust) Fe_2O_3. Modern industrial approaches to forming black oxide include the hot and mid-temperature processes described below. The oxide can also be formed by an electrolytic process in anodizing. They are of interest historically, and are also useful for hobbyists to form black oxide safely with little equipment and without toxic chemicals.

Low temperature oxide, also described below, is not a conversion coating—the low-temperature process does not oxidize the iron, but deposits a copper selenium compound.

Hot Black Oxide

Hot baths of sodium hydroxide, nitrates, and nitrites at 141 °C (286 °F) are used to convert the surface of the material into magnetite (Fe_3O_4). Water must be periodically added to the bath, with proper controls to prevent a steam explosion.

Hot blackening involves dipping the part into various tanks. The workpiece is usually "dipped" by automated part carriers for transportation between tanks. These tanks contain, in order, alkaline cleaner, water, caustic soda at 140.5 °C (284.9 °F) (the blackening compound), and finally the sealant, which is usually oil. The caustic soda bonds chemically to the surface of the metal, creating a porous base layer on the part. Oil is then applied to the heated part, which seals it by "sinking" into the applied porous layer. It is the oil that prevents the corrosion of the workpiece. There are many advantages of blackening, mainly:

- Blackening can be done in large batches (ideal for small parts).

- There is no significant dimensional impact (the blacking process creates a layer about a micrometre thick).

- It is far cheaper than similar corrosion protection systems, such as paint and electroplating.

- The oldest and most widely used specification for hot black oxide is MIL-DTL-13924, which covers four classes of processes for different substrates. Alternate specifications include AMS 2485, ASTM D769, and ISO 11408.

This is the process used to blacken wire ropes for theatrical applications and flying effects.

Mid-temperature Black Oxide

Like hot black oxide, mid-temperature black oxide converts the surface of the metal to magnetite

(Fe_3O_4). However, mid-temperature black oxide blackens at a temperature of 220–245 °F (104–118 °C), significantly less than hot black oxide. This is advantageous because it is below the solution's boiling point, meaning there are no caustic fumes produced.

Since mid-temperature black oxide is most comparable to hot black oxide, it also can meet the military specification MIL-DTL-13924, as well as AMS 2485.

Cold Black Oxide

Cold black oxide is applied at room temperature. It is not an oxide conversion coating, but rather a deposited copper selenium compound. Cold black oxide offers higher productivity and is convenient for in-house blackening. This coating produces a similar color to the one the oxide conversion does, but tends to rub off easily and offers less abrasion resistance. The application of oil, wax, or lacquer brings the corrosion resistance up to par with the hot and mid-temperature. One application for cold black oxide process would be in tooling and architectural finishing on steel (patina for steel). It is also known as cold bluing.

Copper

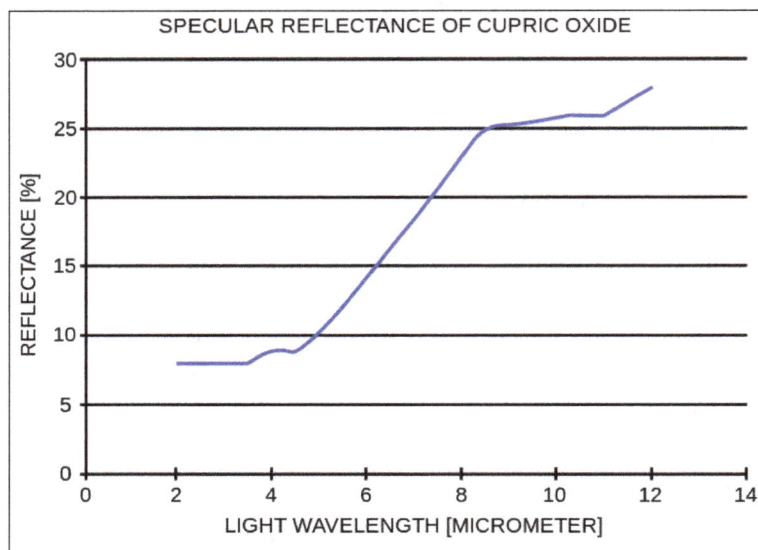

Black oxide for copper, sometimes known by the trade name Ebonol C, converts the copper surface to cupric oxide. For the process to work the surface has to have at least 65% copper; for copper surfaces that have less than 90% copper it must first be pretreated with an activating treatment. The finished coating is chemically stable and very adherent. It is stable up to 400 °F (204 °C); above this temperature the coating degrades due to oxidation of the base copper. To increase corrosion resistance, the surface may be oiled, lacquered, or waxed. It is also used as a pre-treatment for painting or enamelling. The surface finish is usually satin, but it can be turned glossy by coating in a clear high-gloss enamel.

On a microscopic scale dendrites form on the surface finish, which trap light and increase absorptivity. Because of this property the coating is used in aerospace, microscopy and other optical applications to minimise light reflection.

In printed circuit boards (PCBs), the use of black oxide provides better adhesion for the fiberglass laminate layers. The PCB is dipped in a bath containing hydroxide, hypochlorite, and cuprate, which becomes depleted in all three components. This indicates that the black copper oxide comes partially from the cuprate and partially from the PCB copper circuitry. Under microscopic examination, there is no copper(I) oxide layer.

An applicable U.S. military specification is MIL-F-495E.

Stainless Steel

Hot black oxide for stainless steel is a mixture of caustic, oxidizing, and sulfur salts. It blackens 300 and 400 series and the precipitation-hardened 17-4 PH stainless steel alloys. The solution can be used on cast iron and mild low-carbon steel. The resulting finish complies with military specification MIL-DTL–13924D Class 4 and offers abrasion resistance. Black oxide finish is used on surgical instruments in light-intensive environments to reduce eye fatigue.

Room-temperature blackening for stainless steel occurs by auto-catalytic reaction of copper-selenide depositing on the stainless-steel surface. It offers less abrasion resistance and the same corrosion protection as the hot blackening process. One application for room-temperature blackening is in architectural finishes (patina for stainless steel).

Zinc

Black oxide for zinc is also known by the trade name Ebonol Z. Another product is Ultra-Blak 460, which blackens zinc-plated and galvanized surfaces without using any chrome and zinc die-casts.

Bluing (Steel)

Bluing is a passivation process in which steel is partially protected against rust, and is named after the blue-black appearance of the resulting protective finish. True gun bluing is an electrochemical conversion coating resulting from an oxidizing chemical reaction with iron on the surface selectively forming magnetite (Fe_3O_4), the black oxide of iron. Black oxide provides minimal protection against corrosion, unless also treated with a water-displacing oil to reduce wetting and galvanic action. A distinction can be made between traditional bluing and some other more modern black oxide coatings, although bluing is a subset of black oxide coatings.

In comparison, rust, the red oxide of iron (Fe_2O_3), undergoes an extremely large volume change upon hydration; as a result, the oxide easily flakes off causing the typical reddish rusting away of iron. "Cold", "Hot", "Rust Blue" and "Fume Blue" are oxidizing processes simply referred to as bluing.

"Cold" bluing is generally a selenium dioxide based compound that colours steel black, or more often a very dark grey. It is a difficult product to apply evenly, offers minimal protection and is generally best used for small fast repair jobs and touch-ups.

The "Hot" process is an alkali salt solution using potassium nitrite or sodium nitrate and sodium hydroxide, referred to as "Traditional Caustic Black", that is typically done at an elevated temperature, 135 to 155 °C (275 to 311 °F). This method was adopted by larger firearm companies for large scale, more economical bluing. It does provide good rust resistance which is improved with the use of oil.

"Rust Bluing" and "Fume Bluing" provide the best rust and corrosion resistance as the process continually converts any metal that is capable of rusting into magnetite (Fe_3O_4). Treating with an oiled coating enhances the protection offered by the bluing. This process is also the only process safely used to re-blue vintage shotguns. Many double barrelled shotguns are soft soldered (Lead)/ silver brazed together and many of the parts are attached by that method also. The higher temperatures of the other processes as well as their caustic nature will weaken the soldered joints and make the gun hazardous to use.

Bluing can also be done in a furnace, for example for a sword or other item traditionally made by a blacksmith or specialist such as a weaponsmith. Blacksmith products to this day may occasionally be found made from blued steel by traditional craftsmen in cultures and segments of society who use that technology either by necessity or choice.

Bluing is most commonly used by gun manufacturers, gunsmiths, and gun owners to improve the cosmetic appearance of and provide a measure of corrosion resistance to their firearms. It was also used by machinists, who protected and beautified tools made for their own use. Bluing also helps to maintain the metal finish by resisting superficial scratching, and also helps to reduce glare to the eyes of the shooter when looking down the barrel of the gun. All blued parts still need to be properly oiled to prevent rust. Bluing, being a chemical conversion coating, is not as robust against wear and corrosion resistance as plated coatings, and is typically no thicker than 2.5 micrometres (0.0001 inches). For this reason, it is considered not to add any appreciable thickness to precisely-machined gun parts.

New guns are typically available in blued finish options offered as the least-expensive finish, and this finish is also the least effective at providing rust resistance, relative to other finishes such as Parkerizing or hard chrome plating or nitriding processes like Tenifer.

Bluing is also used for providing coloring for steel parts of fine clocks and other fine metalwork. This is often achieved without chemicals by simply heating the steel until a blue oxide film appears. The blue appearance of the oxide film is also used as an indication of temperature when tempering carbon steel after hardening, indicating a state of temper suitable for springs.

Bluing is also used in seasoning cast-iron cookware, to render it relatively rust-proof and non-stick. In this case cooking oil, rather than gun oil, acts to displace water and prevent rust.

Bluing is often a hobbyist endeavor, and there are many methods of bluing, and continuing debates about the relative efficacy of each method.

Historically, razor blades were often blued steel. A non-linear resistance property of the blued steel of razor blades, foreshadowing the same property that would later be discovered in semiconductor diode junctions, along with the ready availability of blued steel razor blades, led to the use of razor blades as a detector in the crystal set AM radios which were often built by soldiers during World War II.

Hot Bluing

Bluing may be applied, for example, by immersing the steel parts to be blued in a solution of potassium nitrate, sodium hydroxide, and water heated to the boiling point, 275 °F to 310 °F (135 °C to 154 °C) depending on the recipe. Similarly, stainless steel parts may be immersed in a mixture of nitrates and chromates, similarly heated. Either of these two methods are called hot bluing. There are many other methods of hot bluing. Hot bluing is the current standard in gun bluing, as both it and rust bluing provide the most permanent degree of rust-resistance and cosmetic protection of exposed gun metal.

Acid solution applied to bare metal.

After boiling rusted parts.

After eight rust, carding and oiling sessions.

Rust Bluing

Rust bluing was developed between hot and cold bluing processes. It was originally used by gunsmiths in the 19th century to blue firearms prior to the development of hot bluing processes. The process was to coat the gun parts in an acid solution, let the parts rust uniformly, then immerse the parts in boiling water to convert the red oxide Fe_2O_3 to black oxide Fe_3O_4, which forms a more protective, stable coating than the red oxide. The boiling water also removes any remaining residue from the applied acid solution (often nitric acid and hydrochloric acid diluted in water). Then loose oxide was carded (scrubbed) off, using a carding brush or wheel. A carding brush is a wire brush with very soft, thin (usually about .002 thick) wires. This process is repeated until the desired depth of color is achieved or the metal simply will not color any further. This is one of the reasons rust and fume bluing tend to be more rust resistant than any other method. The parts are then oiled and allowed to stand overnight. This process leaves a deep blue/black finish.

Modern home hobbyist versions of this process typically use a hydrogen peroxide and salt solution, sometimes with vinegar, for the rusting step to avoid the need for more dangerous acids.

Fume Bluing

Fume bluing is another process similar to rust bluing. Instead of applying the acid solution directly to the metal parts, the parts are placed in a sealed cabinet with a moisture source, a container of nitric acid and a container of hydrochloric acid. The cabinet is then sealed. The mixed fumes of the acids will produce a uniform rust on the surface of the parts (inside and out) in about 12 hours. The parts are then boiled in distilled water, blown dry, then carded, as with rust bluing.

These processes were later abandoned by major firearm manufacturers as it often took parts days to finish completely, and was very labor-intensive. They are still sometimes used by gunsmiths to obtain an authentic finish for a period gun of the time that rust bluing was in vogue, analogous to the use of browning on earlier representative firearm replicas. Rust bluing is also used on shotgun barrels that are soldered to the rib between the barrels, as hot bluing solutions would dissolve the solder during the bluing process.

Large scale industrial hot bluing is often performed using a bluing furnace. This is an alternative method for creating the black oxide coating. In place of using a hot bath (although at a lower temperature) chemically induced method, it is possible through controlling the temperature to heat steel precisely such as to cause the formation of black oxide selectively over the red oxide. It, too, must be oiled to provide any significant rust resistance.

Cold Bluing

There are also methods of cold bluing, which do not require heat. Commercial products are widely sold in small bottles for cold bluing firearms, and these products are primarily used by individual gun owners for implementing small touch-ups to a gun's finish, to prevent a small scratch from becoming a major source of rust on a gun over time. At least one of the cold bluing solutions contains selenium dioxide, to accomplish the bluing. Selenium containing cold bluing solutions work by depositing a coating of copper selenide on the surface. Cold bluing is not particularly resistant to holster wear, nor does it provide a large degree of rust resistance. Often it does provide an adequate cosmetic touch-up of a gun's finish when applied and additionally oiled on a regular basis. However, rust bluing small areas will often match and blend better and wear better than any cold bluing process.

Other Methods of Bluing and Colouring

Niter Bluing

Parts to be niter blued are steel which has been polished and cleaned, then immersed in a bath of molten salts; typically potassium nitrate and sodium nitrate (sometimes with 9.4 grams (0.33 oz) of manganese dioxide per lb of total nitrate). The mixture is heated to 310 to 321 °C (590 to 610 °F) and the parts are suspended in this solution with wire. The parts must be observed constantly for colour change. The cross section and size of parts will affect the outcome of the finish and time it takes to achieve. This method must not be used on critically heat-treated parts such as receivers, slides or springs. It is generally employed on smaller parts such as pins, screws, sights, etc. The colours will range through straw, gold, brown, purple, blue, teal, then black. Examples of this finish can be seen commonly on older pocket watches whose hands exhibit what is called "peacock blue", a rich iridescent blue.

Niter and colour case.

Colour Case Hardening

This is the predecessor of all metal colouring typically employed in the firearms industry. Contemporary heat-treatable steels did not exist or were in their infancy. Soft, low-carbon steel was used, but strong materials were needed for the receivers of firearms. Initially case hardening was used but did not offer any aesthetics. Colour case hardening occurs when soft steels were packed in a reasonably airtight crucible in a mixture of charred leather, bone charcoal and wood charcoal. This crucible was heated to 730 °C (1,350 °F) for up to 6 hours (the longer the heat was applied the thicker the case hardening). At the end of this heating process the crucible is removed from the oven and positioned over a bath of water with air forced through a perforated coil in the bottom of the bath. The bottom of the crucible is opened allowing the contents to drop into the rapidly bubbling water. The differential cooling causes patterns of colours to appear as well as hardening the part.

Different colours can be achieved through variations of this method including quenching in oil instead of water.

Browning

Browning is controlled red rust Fe_2O_3 and is also known as pluming or plum brown. One can generally use the same solution to brown as to blue. The difference is immersion in boiling water for bluing. The rust then turns to black-blue Fe_3O_4. Many older browning and bluing formulas are based on corrosive solutions (necessary to cause metal to rust), and often contain cyanide or mercury salts solutions that are especially toxic to humans.

Limitations

Bluing only works on steel, cast iron, or stainless steel parts for protecting against corrosion because it changes iron into Fe_3O_4; it does not work on non-ferrous material. Aluminium and polymer parts cannot be blued, and no corrosion protection is provided. However, the chemicals from the blueing process can accomplish uneven staining on aluminium and polymer parts. Hot bluing should never be attempted on aluminium, as it will react violently with the bath of caustic salts, potentially causing severe chemical burns.

Friction, as from holster wear, will quickly remove cold bluing, and will also remove hot bluing, rust, or fume bluing over long periods of use. It is usually inadvisable to use cold bluing as a touch-up where friction is present. If cold-bluing is the only practical option, the area should be kept oiled to extend the life of the coating as much as possible.

Chrome Plating

Decorative chrome plating on a motorcycle.

Chrome plating (less commonly chromium plating), often referred to simply as chrome, is a technique of electroplating a thin layer of chromium onto a metal object. The chromed layer can be decorative, provide corrosion resistance, ease cleaning procedures, or increase surface hardness. Sometimes, a less expensive imitator of chrome may be used for aesthetic purposes.

Process

Chrome plating a component typically includes these stages:

- Degreasing to remove heavy soiling.

- Manual cleaning to remove all residual traces of dirt and surface impurities.

- Various pretreatments depending on the substrate.

- Placement into the chrome plating vat, where it is allowed to warm to solution temperature.

- Application of plating current for the required time to attain the desired thickness.

There are many variations to this process, depending on the type of substrate being plated. Different substrates need different etching solutions, such as hydrochloric, hydrofluoric, and sulfuric acids. Ferric chloride is also popular for the etching of nimonic alloys. Sometimes the component enters the chrome plating vat while electrically live. Sometimes the component has a conforming anode made from lead/tin or platinized titanium. A typical hard chrome vat plates at about 1 mil (25 μm) per hour.

Various finishing and buffing processes are used in preparing components for decorative chrome plating. The chrome plating chemicals are very toxic. Disposal of chemicals is regulated in most countries.

Some common industry specifications governing the chrome plating process are AMS 2460, AMS 2406, and MIL-STD-1501.

Hexavalent Chromium

Hexavalent chromium plating, also known as hex-chrome, Cr^{6+}, and chrome (VI) plating, uses chromium trioxide (also known as chromic anhydride) as the main ingredient. Hexavalent chromium plating solution is used for decorative and hard plating, along with bright dipping of copper alloys, chromic acid anodizing, and chromate conversion coating.

A typical hexavalent chromium plating process is: (1) activation bath, (2) chromium bath, (3) rinse, and (4) rinse. The activation bath is typically a tank of chromic acid with a reverse current run through it. This etches the work-piece surface and removes any scale. In some cases the activation step is done in the chromium bath. The chromium bath is a mixture of chromium trioxide (CrO_3) and sulfuric acid (sulfate, SO_4), the ratio of which varies greatly between 75:1 to 250:1 by weight. This results in an extremely acidic bath (pH 0). The temperature and current density in the bath affect the brightness and final coverage. For decorative coating the temperature ranges from 35 to 45 °C (100 to 110 °F), but for hard coating it ranges from 50 to 65 °C (120 to 150 °F). Temperature is also dependent on the current density, because a higher current density requires a higher temperature. Finally, the whole bath is agitated to keep the temperature steady and achieve a uniform deposition.

Disadvantages

One functional disadvantage of hexavalent chromium plating is low cathode efficiency, which results in bad throwing power. This means it leaves a non-uniform coating, with more on edges and less in inside corners and holes. To overcome this problem the part may be over-plated and ground to size, or auxiliary anodes may be used around the hard-to-plate areas.

From a health standpoint, hexavalent chromium is the most toxic form of chromium. In the U.S., the Environmental Protection Agency regulates it heavily. The EPA lists hexavalent chromium as a hazardous air pollutant because it is a human carcinogen, a "priority pollutant" under the Clean Water Act, and a "hazardous constituent" under the Resource Conservation and Recovery Act. Due to its low cathodic efficiency and high solution viscosity, a toxic mist of water and hexavalent chromium is released from the bath. Wet scrubbers are used to control these emissions. The discharge from the wet scrubbers is treated to precipitate the chromium from the solution because it cannot remain in the waste water.

Maintaining a bath surface tension less than 35 dynes/cm requires a frequent cycle of treating the bath with a wetting agent and confirming the effect on surface tension. Traditionally, surface tension is measured with a stalagmometer. This method is, however, tedious and suffers from inaccuracy (errors up to 22 dynes/cm have been reported), and is dependent on the user's experience and capabilities.

Additional toxic waste created from hexavalent chromium baths include lead chromates, which form in the bath because lead anodes are used. Barium is also used to control the sulfate concentration, which leads to the formation of barium sulfate ($BaSO_4$).

Trivalent Chromium

Trivalent chromium plating, also known as tri-chrome, Cr^{3+}, and chrome (III) plating, uses chromium sulfate or chromium chloride as the main ingredient. Trivalent chromium plating is an alternative to hexavalent chromium in certain applications and thicknesses (e.g. decorative plating).

A trivalent chromium plating process is similar to the hexavalent chromium plating process, except for the bath chemistry and anode composition. There are three main types of trivalent chromium bath configurations:

- A chloride- or sulfate-based electrolyte bath using graphite or composite anodes, plus additives to prevent the oxidation of trivalent chromium to the anodes.

- A sulfate-based bath that uses lead anodes surrounded by boxes filled with sulfuric acid (known as shielded anodes), which keeps the trivalent chromium from oxidizing at the anodes.

- A sulfate-based bath that uses insoluble catalytic anodes, which maintains an electrode potential that prevents oxidation.

The trivalent chromium-plating process can plate the workpieces at a similar temperature, rate and hardness, as compared to hexavalent chromium. Plating thickness ranges from 0.005 to 0.05 mils (0.13 to 1.27 μm).

Advantages and Disadvantages

The functional advantages of trivalent chromium are higher cathode efficiency and better throwing power. Better throwing power means better production rates. Less energy is required because of the lower current densities required. The process is more robust than hexavalent chromium because it can withstand current interruptions.

From a health standpoint, trivalent chromium is intrinsically less toxic than hexavalent chromium. Because of the lower toxicity it is not regulated as strictly, which reduces overhead costs. Other health advantages include higher cathode efficiencies, which lead to less chromium air emissions; lower concentration levels, resulting in less chromium waste and anodes that do not decompose.

One of the disadvantages when the process was first introduced was that decorative customers disapproved of the color differences. Companies now use additives to adjust the color. In hard coating applications, the corrosion resistance of thicker coatings is not quite as good as it is with hexavalent chromium. The cost of the chemicals is greater, but this is usually offset by greater production rates and lower overhead costs. In general, the process must be controlled more closely than in hexavalent chromium plating, especially with respect to metallic impurities. This means processes that are hard to control, such as barrel plating, are much more difficult using a trivalent chromium bath.

Types

Decorative

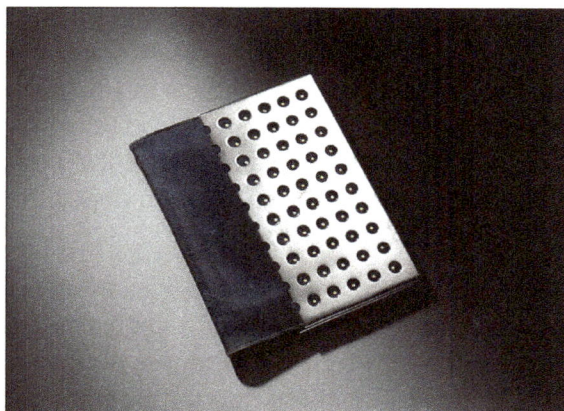
Art Deco portfolio with chrome-plated cover, ca 1925.

Decorative chrome is designed to be aesthetically pleasing and durable. Thicknesses range from 0.002 to 0.02 mils (0.05 to 0.5 μm), however they are usually between 0.005 and 0.01 mils (0.13 and 0.25 μm). The chromium plating is usually applied over bright nickel plating. Typical base materials include steel, aluminium, plastic, copper alloys, and zinc alloys. Decorative chrome plating is also very corrosion resistant and is often used on car parts, tools and kitchen utensils.

Hard

Hard chrome plating.

Hard chrome, also known as industrial chrome or engineered chrome, is used to reduce friction, improve durability through abrasion tolerance and wear resistance in general, minimize galling or seizing of parts, expand chemical inertness to include a broader set of conditions (especially oxidation resistance, arguably its most famous quality), and bulking material for worn parts to restore their original dimensions. It is very hard, measuring between 65 and 69 HRC (also based on the base metal's hardness). Hard chrome tends to be thicker than decorative chrome, with standard thicknesses in nonsalvage applications ranging from 0.02 to 0.04 mm (20 to 40 μm), but it can be an order of magnitude thicker for extreme wear resistance requirements, in such cases 0.1 mm (100 μm) or thicker provides optimal results. Unfortunately, such thicknesses emphasize the limitations of the process, which are overcome by plating extra thickness then grinding down and lapping to meet requirements or to improve the overall aesthetics of the "chromed" piece. Increasing plating thickness amplifies surface defects and roughness in proportional severity, because hard chrome does not have a leveling effect. Pieces that are not ideally shaped in reference to electric field geometries (nearly every piece sent in for plating, except spheres and egg shaped objects) require even thicker plating to compensate for non-uniform deposition, and much of it is wasted when grinding the piece back to desired dimensions.

Modern "engineered coatings" do not suffer such drawbacks, which often price hard chrome out due to labor costs alone. Hard chrome replacement technologies outperform hard chrome in wear resistance, corrosion resistance, and cost. Rockwell hardness 80 is not extraordinary for such materials. Using spray deposition, uniform thickness that often requires no further polishing or machining is a standard feature of modern engineered coatings. These coatings are often composites of polymers, metals, and ceramic powders or fibers as proprietary embodiments protected by patents or as trade secrets, and thus are usually known by brand names.

Hard chromium plating is subject to different types of quality requirements depending on the application; for instance, the plating on hydraulic piston rods are tested for corrosion resistance with a salt spray test.

Automotive Use

Most bright decorative items affixed to cars are referred to as "chrome", meaning steel that has undergone several plating processes to endure the temperature changes and weather that a car is subject to outdoors. Triple plating is the most expensive and durable process, which involves plating the steel first with copper and then nickel before the chromium plating is applied.

Prior to the application of chrome in the 1920s, nickel electroplating was used. In the short production run prior to the US entry into the Second World War, the government banned plating to save chromium and automobile manufacturers painted the decorative pieces in a complementary color. In the last years of the Korean War, the US contemplated banning chrome in favor of several cheaper processes (such as plating with zinc and then coating with shiny plastic).

In 2007, a Restriction of Hazardous Substances Directive (RoHS) was issued banning several toxic substances for use in the automotive industry in Europe, including hexavalent chromium, which is used in chrome plating. However, chrome plating is metal and contains no hexavalent chromium after it is rinsed, so chrome plating is not banned.

APPLICATIONS OF ELECTROCHEMICAL REACTION

Applications

Electrochemical processes are used in many ways and their use is likely to increase because they can replace polluting chemical situations with nonpolluting electrochemical ones. In many fields, however, applications have been profitable for some time. Major categories are listed below:

Metallurgy

All technologically important metals, except iron and steel, are either obtained or refined by electrochemical processes; for example, aluminum, titanium, alkaline earth, and alkali metals are obtained by electrodeposition from molten salts, and copper is refined by electrolysis in aqueous copper sulfate solutions.

Electroplating

One of the major ways of both decorating objects and improving their resistance to corrosionis by electroplating them. All major metal-working industries, particularly the automobile industry, have large electroplating plants.

Chemical Industry

Electrolysis of brine to obtain chlorine and caustic soda is an electrochemical process that has become one of the largest volume productions in the chemical industry. Modern processes cover a wide field, from the production of a variety of inorganic compounds to the production of such synthetic fibres as nylon. Intensive research in organic electrochemistry promises major developments in application, particularly with the prospect of greatly reduced electricity costs expected eventually to arise from the development of controlled fusion.

Batteries

Electrochemical storage of electricity is effected in batteries. Such devices are electrochemical cells and consist of two electrodes per unit. As the electricity to be stored is accepted on the plates of the cell, it converts substances on the plates to new substances having a higher energy than the old ones. When it is desired to make the electricity available again, the terminals of the battery are connected to the load and the substances on the battery plates retransform themselves to those originally present, giving off electricity as a product of their electrochemical reactions. The steadily rising production of the lead-acid battery is largely the result of its use for starting the internal-combustion engine, which has had an equally steady rise. Other electrochemical systems are also used as storers. The nickel-iron (Edison cell) and nickel-cadmium battery with alkaline electrolyte are both used in applications where longer lives than those of the lead-acid battery are needed; the silver-zinc battery is used to start airplane engines because of its high power per unit of weight. A variety of new systems is being investigated for covering other needs. One of the greatest challenges to electrochemists and electrochemical engineers is that of producing a battery with sufficient power and energy density to run an automobile the way gasoline (petrol) does. Even if the best hypothetical predictions for removal of polluting chemicals from automobile exhausts is realized, the cleanup will not be sufficient because the expected growth of the automobile population will continue to increase the pollutant rate.

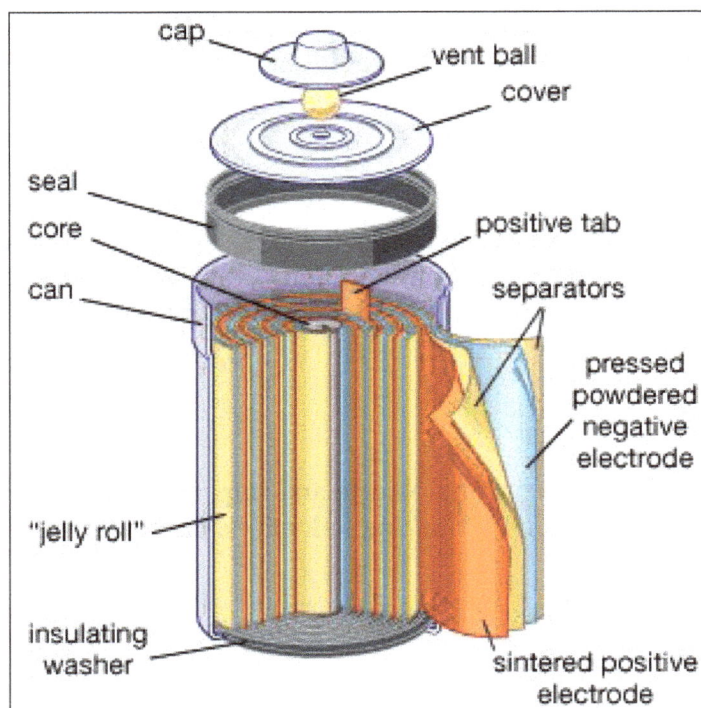

Nickel (hydroxide)-cadmium cell of "jelly roll" construction. This rechargable battery is commonly used in portable devices.

Fuel Cells

The energy of chemical reactions is converted into electrical energy in fuel cells. In these, the fuel (e.g., hydrogen, hydrazine) is fed continuously to one electrode, while oxygen from the air is reacting at the other one. The efficiency of energy conversion in fuel cells is more than twice that attainable by conventional means—for example, by means of internal combustion.

Analytical Chemistry

In analytical chemistry, most modern automated instrumental analysis is based on electrode processes—for example, potentiometry, used to measure ionization constant.

Biological Research

In biology the idea that many biological processes, from blood clotting to the transfer of nerve impulses, are electrochemical in nature continues to spread. The biological conversion of the chemical energy of food to mechanical energy takes place at an efficiency so high that it is difficult to explain without electrochemical mechanisms. Intensive research is developing in various directions in bioelectrochemistry.

References

- Robertson, William (2010). More Chemistry Basics. National Science Teachers Association. P. 82. ISBN 978-1-936137-74-9

- Electrochemical-reaction, science: britannica.com, Retrieved 29 March, 2019

- Bard, Allen J.; Faulkner, Larry R. (January 2001). Electrochemical methods: fundamentals and applications. New York: Wiley. ISBN 978-0-471-04372-0. Retrieved 27 February 2009

- Geiger, William E. (2007-11-01). "Organometallic Electrochemistry: Origins, Development, and Future". Organometallics. 26 (24): 5738–5765. Doi:10.1021/om700558k

- "redox – definition of redox in English | Oxford Dictionaries". Oxford Dictionaries | English. Archived from the original on 2017-10-01. Retrieved 2017-05-15

- Vanloon, Gary; Duffy, Stephen (2011). Environmental Chemistry -(*Gary Wallace) a global perspective (3rd ed.). Oxford University Press. Pp. 235–248. ISBN 978-0-19-922886-7

- Applications, electrochemical-reaction, science: britannica.com, Retrieved 16 January, 2019

Electrochemical Engineering

The domain within chemical engineering which focuses on the technological applications of electrochemical phenomena is referred to as electrochemical engineering. Electrochemical gas sensor, glow battery, electrophoresis, etc. are some of the concepts that fall in its domain. This chapter discusses in detail these concepts related to electrochemical engineering.

Electrochemical engineering is the knowledge required to either design and run an industrial plant which includes an electrolytic stage for the production of chemicals or to produce an electrolytic device for the generation of power. The former involves the use of electric power for the production of chemicals and the latter the use of chemicals for the production of electric power. There are also a number of processes based on electrolysis, for example, electrochemical machining and electrophoretic painting, which clearly belong to the domain of electrochemical engineering. Nonelectrolytic processes such as gas discharge phenomena can justifiably be taken as outside the realm of the definition.

Electrochemical Energy Conversion

Electrochemical cells and systems play a key role in a wide range of industry sectors. These devices are critical enabling technologies for renewable energy; energy management, conservation, and storage; pollution control/monitoring; and greenhouse gas reduction. A large number of electrochemical energy technologies have been developed in the past. These systems continue to be optimized in terms of cost, life time, and performance, leading to their continued expansion into existing and emerging market sectors. The more established technologies such as deep-cycle batteries and sensors are being joined by emerging technologies such as fuel cells, large format lithium-ion batteries, electrochemical reactors; ion transport membranes and supercapacitors. This growing demand (multi billion dollars) for electrochemical energy systems along with the increasing maturity of a number of technologies is having a significant effect on the global research and development effort which is increasing in both in size and depth. A number of new technologies, which will have substantial impact on the environment and the way we produce and utilize energy, are under development.

ELECTROCHEMICAL REDUCTION OF CARBON DIOXIDE

The electrochemical reduction of carbon dioxide (ERC) is the conversion of carbon dioxide to more reduced chemical species using electrical energy. The first examples of electrochemical reduction of carbon dioxide are from the 19th century, when carbon dioxide was reduced to carbon monoxide using a zinc cathode. Research in this field intensified in the 1980s following the oil embargoes of the 1970s. Electrochemical reduction of carbon dioxide represents a possible means of producing chemicals or fuels, converting carbon dioxide (CO_2) to organic feedstocks such as formic acid (HCOOH), methanol (CH_3OH), ethylene (C_2H_4), methane (CH_4), and carbon monoxide (CO). In 2018, researchers from the University of Delaware reported a general techno-economic analysis of CO_2 electrolysis technology.

Chemicals from Carbon Dioxide

In carbon fixation, plants convert carbon dioxide into sugars, from which many biosynthetic pathways originate. The catalyst responsible for this conversion, RuBisCo, is the most common protein on earth. Some anaerobic organisms employ enzymes to convert CO_2 to carbon monoxide, from which fatty acids can be made.

In industry, a few products are made from CO_2, including urea, salicylic acid, methanol, and certain inorganic and organic carbonates. In the laboratory, carbon dioxide is sometimes used to prepare carboxylic acids. No electrochemical process involving CO_2 has been commercialized.

Electrocatalysis

The electrochemical reduction of carbon dioxide to CO is usually described as:

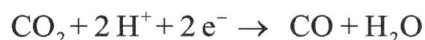

$$CO_2 + 2\,H^+ + 2\,e^- \rightarrow CO + H_2O$$

The redox potential for this reaction is similar to that for hydrogen evolution in aqueous electrolytes, thus electrochemical reduction of CO_2 is usually competitive with hydrogen evolution reaction.

Electrochemical methods have gained significant attention: 1) at ambient pressure and room temperature; 2) in connection with renewable energy sources 3) competitive controllability, modularity and scale-up are relatively simple. The electrochemical reduction or electrocatalytic conversion of CO_2 can produce value-added chemicals such methane, ethylene, ethane, etc., and the products are mainly dependent on the selected catalysts and operating potentials (applying reduction voltage).

Although an electrochemical route to CO (or other chemicals) has not been commercialized, a variety of homogeneous and heterogeneous catalysts have been evaluated. Many such processes are assumed to operate via the intermediacy of metal carbon dioxide complexes. Generally speaking, the processes developed up to 2010 either had poor thermodynamic efficiency (high overpotential), low current efficiency, low selectivity, slow kinetics, and poor stability. In 2011, workers from Dioxide Materials and University of Illinois showed that the combination of two catalysts could eliminate the high overpotential More recently, the same group showed that the process was stable for 6 months at over 90% selectivity. Studies have shown that a gas-diffusion electrode design could promote the reaction rate of electrochemical CO_2 reduction to CO and multi-carbon products.

FUEL CELL

Demonstration model of a direct-methanol fuel cell. The actual fuel cell stack is the layered cube shape in the center of the image.

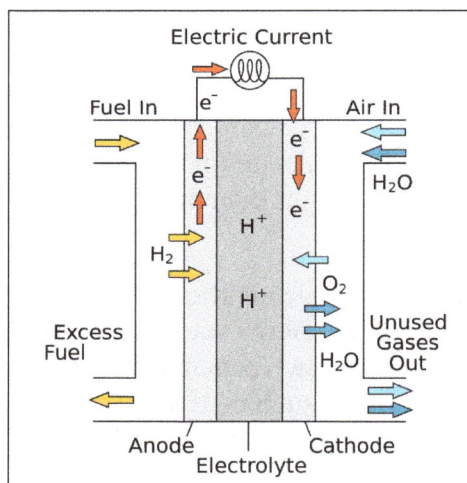

Scheme of a proton-conducting fuel cell.

A fuel cell is an electrochemical cell that converts the chemical energy of a fuel (often hydrogen) and an oxidizing agent (often oxygen) into electricity through a pair of redox reactions. Fuel cells are different from most batteries in requiring a continuous source of fuel and oxygen (usually from air) to sustain the chemical reaction, whereas in a battery the chemical energy usually comes from metals and their ions or oxides that are commonly already present in the battery, except in flow batteries. Fuel cells can produce electricity continuously for as long as fuel and oxygen are supplied.

The first fuel cells were invented by Sir William Grove in 1838. The first commercial use of fuel cells came more than a century later following the invention of the hydrogen–oxygen fuel cell by Francis Thomas Bacon in 1932. The alkaline fuel cell, also known as the Bacon fuel cell after its inventor, has been used in NASA space programs since the mid-1960s to generate power for satellites and space capsules. Since then, fuel cells have been used in many other applications. Fuel cells are used for primary and backup power for commercial, industrial and residential buildings and in remote or inaccessible areas. They are also used to power fuel cell vehicles, including forklifts, automobiles, buses, boats, motorcycles and submarines.

There are many types of fuel cells, but they all consist of an anode, a cathode, and an electrolyte that allows ions, often positively charged hydrogen ions (protons), to move between the two sides of the fuel cell. At the anode a catalyst causes the fuel to undergo oxidation reactions that generate ions (often positively charged hydrogen ions) and electrons. The ions move from the anode to the cathode through the electrolyte. At the same time, electrons flow from the anode to the cathode through an external circuit, producing direct current electricity. At the cathode, another catalyst causes ions, electrons, and oxygen to react, forming water and possibly other products. Fuel cells are classified by the type of electrolyte they use and by the difference in startup time ranging from 1 second for proton exchange membrane fuel cells (PEM fuel cells, or PEMFC) to 10 minutes for solid oxide fuel cells (SOFC). A related technology is flow batteries, in which the fuel can be regenerated by recharging. Individual fuel cells produce relatively small electrical potentials, about 0.7 volts, so cells are "stacked", or placed in series, to create sufficient voltage to meet an application's requirements. In addition to electricity, fuel cells produce water, heat and, depending on the fuel source, very small amounts of nitrogen dioxide and other emissions. The energy efficiency of a fuel cell is generally between 40–60%; however, if waste heat is captured in a cogeneration scheme, efficiencies of up to 85% can be obtained.

The fuel cell market is growing, and in 2013 Pike Research estimated that the stationary fuel cell market will reach 50 GW by 2020.

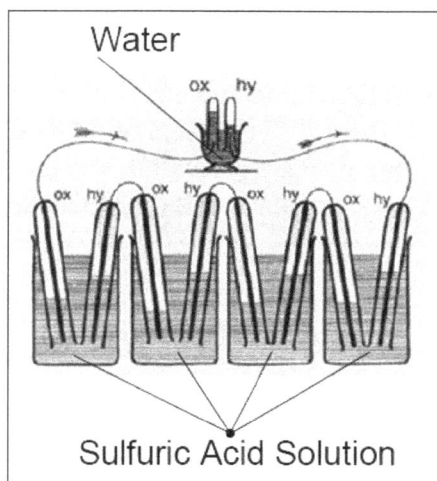

Sketch of Sir William Grove's 1839 fuel cell.

He used a combination of sheet iron, copper and porcelain plates, and a solution of sulphate of copper and dilute acid. In a letter to the same publication written in December 1838 but published in June 1839, German physicist Christian Friedrich Schönbein discussed the first crude fuel cell that he had invented. His letter discussed current generated from hydrogen and oxygen dissolved in water. Grove later sketched his design, in 1842, in the same journal. The fuel cell he made used similar materials to today's phosphoric acid fuel cell.

In 1932, English engineer Francis Thomas Bacon successfully developed a 5 kW stationary fuel cell. The alkaline fuel cell (AFC), also known as the Bacon fuel cell after its inventor, is one of the most developed fuel cell technologies, which NASA has used since the mid-1960s.

In 1955, W. Thomas Grubb, a chemist working for the General Electric Company (GE), further modified the original fuel cell design by using a sulphonated polystyrene ion-exchange membrane as the electrolyte. Three years later another GE chemist, Leonard Niedrach, devised a way of depositing platinum onto the membrane, which served as catalyst for the necessary hydrogen oxidation and oxygen reduction reactions. This became known as the "Grubb-Niedrach fuel cell". GE went on to develop this technology with NASA and McDonnell Aircraft, leading to its use during Project Gemini. This was the first commercial use of a fuel cell. In 1959, a team led by Harry Ihrig built a 15 kW fuel cell tractor for Allis-Chalmers, which was demonstrated across the U.S. at state fairs. This system used potassium hydroxide as the electrolyte and compressed hydrogen and oxygen as the reactants. Later in 1959, Bacon and his colleagues demonstrated a practical five-kilowatt unit capable of powering a welding machine. In the 1960s, Pratt & Whitney licensed Bacon's U.S. patents for use in the U.S. space program to supply electricity and drinking water (hydrogen and oxygen being readily available from the spacecraft tanks). In 1991, the first hydrogen fuel cell automobile was developed by Roger Billings.

UTC Power was the first company to manufacture and commercialize a large, stationary fuel cell system for use as a co-generation power plant in hospitals, universities and large office buildings.

In recognition of the fuel cell industry and America's role in fuel cell development, the US Senate recognized 8 October 2015 as National Hydrogen and Fuel Cell Day, passing S. RES 217. The date was chosen in recognition of the atomic weight of hydrogen (1.008).

Types of Fuel Cells: Design

Fuel cells come in many varieties; however, they all work in the same general manner. They are made up of three adjacent segments: the anode, the electrolyte, and the cathode. Two chemical reactions occur at the interfaces of the three different segments. The net result of the two reactions is that fuel is consumed, water or carbon dioxide is created, and an electric current is created, which can be used to power electrical devices, normally referred to as the load.

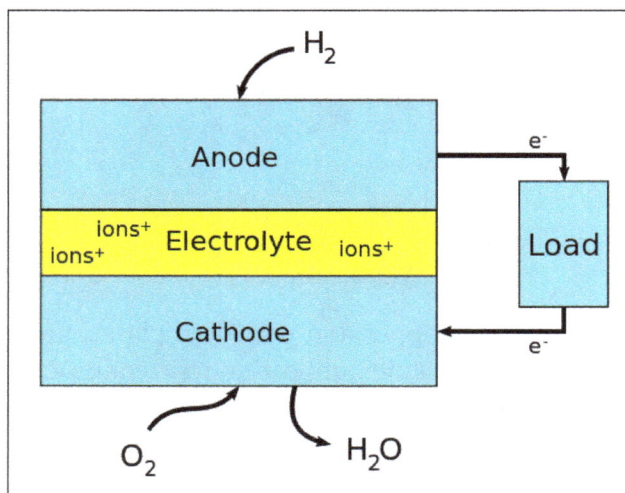

A block diagram of a fuel cell.

At the anode a catalyst oxidizes the fuel, usually hydrogen, turning the fuel into a positively charged ion and a negatively charged electron. The electrolyte is a substance specifically designed so ions can pass through it, but the electrons cannot. The freed electrons travel through a wire creating the electric current. The ions travel through the electrolyte to the cathode. Once reaching the cathode, the ions are reunited with the electrons and the two react with a third chemical, usually oxygen, to create water or carbon dioxide.

Design features in a fuel cell include:

- The electrolyte substance, which usually defines the type of fuel cell, and can be made from a number of substances like potassium hydroxide, salt carbonates, and phosphoric acid.

- The fuel that is used. The most common fuel is hydrogen.

- The anode catalyst, usually fine platinum powder, breaks down the fuel into electrons and ions.

- The cathode catalyst, often nickel, converts ions into waste chemicals, with water being the most common type of waste.

- Gas diffusion layers that are designed to resist oxidization.

A typical fuel cell produces a voltage from 0.6 V to 0.7 V at full rated load. Voltage decreases as current increases, due to several factors:

- Activation loss.

- Ohmic loss (voltage drop due to resistance of the cell components and interconnections).

- Mass transport loss (depletion of reactants at catalyst sites under high loads, causing rapid loss of voltage).

To deliver the desired amount of energy, the fuel cells can be combined in series to yield higher voltage, and in parallel to allow a higher current to be supplied. Such a design is called a fuel cell stack. The cell surface area can also be increased, to allow higher current from each cell. Within the stack, reactant gases must be distributed uniformly over each of the cells to maximize the power output.

Proton-exchange Membrane Fuel Cells (PEMFCs)

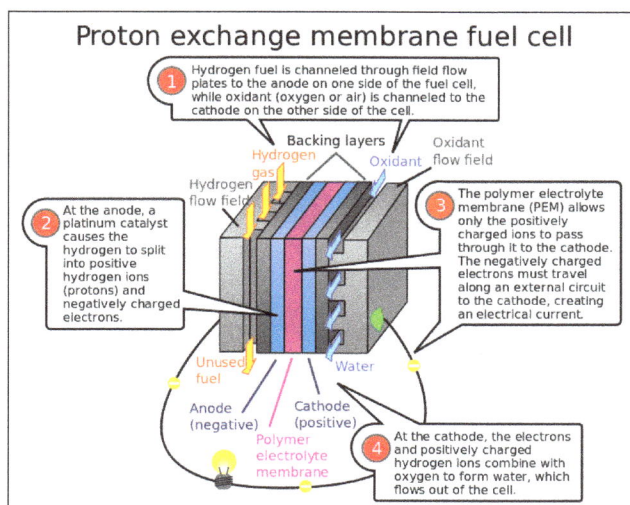

Proton exchange membrane fuel cell

1. Hydrogen fuel is channeled through field flow plates to the anode on one side of the fuel cell, while oxidant (oxygen or air) is channeled to the cathode on the other side of the cell.

2. At the anode, a platinum catalyst causes the hydrogen to split into positive hydrogen ions (protons) and negatively charged electrons.

3. The polymer electrolyte membrane (PEM) allows only the positively charged ions to pass through it to the cathode. The negatively charged electrons must travel along an external circuit to the cathode, creating an electrical current.

4. At the cathode, the electrons and positively charged hydrogen ions combine with oxygen to form water, which flows out of the cell.

Construction of a high-temperature PEMFC: Bipolar plate as electrode with in-milled gas channel structure, fabricated from conductive composites (enhanced with graphite, carbon black, carbon fiber, and carbon nanotubes for more conductivity); Porous carbon papers; reactive layer, usually on the polymer membrane applied; polymer membrane.

Condensation of water produced by a PEMFC on the air channel wall.
The gold wire around the cell ensures the collection of electric current.

In the archetypical hydrogen–oxide proton-exchange membrane fuel cell design, a proton-conducting polymer membrane (typically nafion) contains the electrolyte solution that separates the anode and cathode sides. This was called a "solid polymer electrolyte fuel cell" (SPEFC) in the early 1970s, before the proton exchange mechanism was well understood. (Notice that the synonyms "polymer electrolyte membrane" and "proton exchange mechanism" result in the same acronym).

On the anode side, hydrogen diffuses to the anode catalyst where it later dissociates into protons and electrons. These protons often react with oxidants causing them to become what are commonly referred to as multi-facilitated proton membranes. The protons are conducted through the membrane to the cathode, but the electrons are forced to travel in an external circuit (supplying power) because the membrane is electrically insulating. On the cathode catalyst, oxygen molecules react with the electrons (which have traveled through the external circuit) and protons to form water.

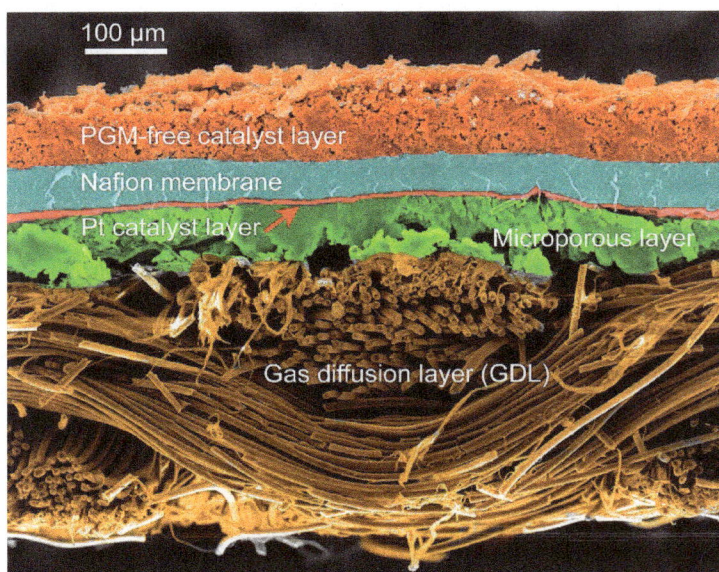

SEM micrograph of a PEMFC MEA cross-section with a non-precious metal catalyst cathode and

Pt/C anode. False colors applied for clarity. In addition to this pure hydrogen type, there are hydrocarbon fuels for fuel cells, including diesel, methanol and chemical hydrides. The waste products with these types of fuel are carbon dioxide and water. When hydrogen is used, the CO_2 is released when methane from natural gas is combined with steam, in a process called steam methane reforming, to produce the hydrogen. This can take place in a different location to the fuel cell, potentially allowing the hydrogen fuel cell to be used indoors—for example, in fork lifts.

The different components of a PEMFC are:

- bipolar plates.

- electrodes.

- catalyst.

- membrane.

- the necessary hardware such as current collectors and gaskets.

The materials used for different parts of the fuel cells differ by type. The bipolar plates may be made of different types of materials, such as, metal, coated metal, graphite, flexible graphite, C–C composite, carbon–polymer composites etc. The membrane electrode assembly (MEA) is referred as the heart of the PEMFC and is usually made of a proton exchange membrane sandwiched between two catalyst-coated carbon papers. Platinum and similar type of noble metals are usually used as the catalyst for PEMFC. The electrolyte could be a polymer membrane.

Proton Exchange Membrane Fuel Cell Design Issues

- Cost: In 2013, the Department of Energy estimated that 80-kW automotive fuel cell system costs of US$67 per kilowatt could be achieved, assuming volume production of 100,000 automotive units per year and US$55 per kilowatt could be achieved, assuming volume production of 500,000 units per year. Many companies are working on techniques to reduce cost in a variety of ways including reducing the amount of platinum needed in each individual cell. Ballard Power Systems has experimented with a catalyst enhanced with carbon silk, which allows a 30% reduction (1 mg/cm² to 0.7 mg/cm²) in platinum usage without reduction in performance. Monash University, Melbourne uses PEDOT as a cathode. A 2011 published study doi: 10.1021/ja1112904 documented the first metal-free electrocatalyst using relatively inexpensive doped carbon nanotubes, which are less than 1% the cost of platinum and are of equal or superior performance. A recently published report demonstrated how the environmental burdens change when using carbon nanotubes as carbon substrate for platinum.

- Water and air management (in PEMFCs): In this type of fuel cell, the membrane must be hydrated, requiring water to be evaporated at precisely the same rate that it is produced. If water is evaporated too quickly, the membrane dries, resistance across it increases, and eventually it will crack, creating a gas "short circuit" where hydrogen and oxygen combine directly, generating heat that will damage the fuel cell. If the water is evaporated too slowly, the electrodes will flood, preventing the reactants from reaching the catalyst and stopping the reaction. Methods to manage water in cells are being developed like electroosmotic pumps focusing on flow control. Just as in a combustion engine, a steady ratio between the reactant and oxygen is necessary to keep the fuel cell operating efficiently.

- Temperature management: The same temperature must be maintained throughout the cell in order to prevent destruction of the cell through thermal loading. This is particularly challenging as the $2H_2 + O_2 \rightarrow 2H_2O$ reaction is highly exothermic, so a large quantity of heat is generated within the fuel cell.

- Durability, service life, and special requirements for some type of cells: Stationary fuel cell applications typically require more than 40,000 hours of reliable operation at a temperature of −35 °C to 40 °C (−31 °F to 104 °F), while automotive fuel cells require a 5,000-hour lifespan (the equivalent of 240,000 km (150,000 mi)) under extreme temperatures. Current service life is 2,500 hours (about 75,000 miles). Automotive engines must also be able to start reliably at −30 °C (−22 °F) and have a high power-to-volume ratio (typically 2.5 kW per liter).

- Limited carbon monoxide tolerance of some (non-PEDOT) cathodes.

Phosphoric Acid Fuel Cell (PAFC)

Phosphoric acid fuel cells (PAFC) were first designed and introduced in 1961 by G. V. Elmore and H. A. Tanner. In these cells phosphoric acid is used as a non-conductive electrolyte to pass positive hydrogen ions from the anode to the cathode. These cells commonly work in temperatures of 150 to 200 degrees Celsius. This high temperature will cause heat and energy loss if the heat is not removed and used properly. This heat can be used to produce steam for air conditioning systems or any other thermal energy consuming system. Using this heat in cogeneration can enhance the efficiency of phosphoric acid fuel cells from 40–50% to about 80%. Phosphoric acid, the electrolyte used in PAFCs, is a non-conductive liquid acid which forces electrons to travel from anode to cathode through an external electrical circuit. Since the hydrogen ion production rate on the anode is small, platinum is used as catalyst to increase this ionization rate. A key disadvantage of these cells is the use of an acidic electrolyte. This increases the corrosion or oxidation of components exposed to phosphoric acid.

Solid Acid Fuel Cell (SAFC)

Solid acid fuel cells (SAFCs) are characterized by the use of a solid acid material as the electrolyte. At low temperatures, solid acids have an ordered molecular structure like most salts. At warmer temperatures (between 140 and 150 degrees Celsius for $CsHSO_4$), some solid acids undergo a phase transition to become highly disordered "superprotonic" structures, which increases conductivity by several orders of magnitude. The first proof-of-concept SAFCs were developed in 2000 using cesium hydrogen sulfate ($CsHSO_4$). Current SAFC systems use cesium dihydrogen phosphate (CsH_2PO_4) and have demonstrated lifetimes in the thousands of hours.

Alkaline Fuel Cell (AFC)

The alkaline fuel cell or hydrogen-oxygen fuel cell was designed and first demonstrated publicly by Francis Thomas Bacon in 1959. It was used as a primary source of electrical energy in the Apollo space program. The cell consists of two porous carbon electrodes impregnated with a suitable catalyst such as Pt, Ag, CoO, etc. The space between the two electrodes is filled with a concentrated solution of KOH or NaOH which serves as an electrolyte. H_2 gas and O_2 gas are bubbled into the electrolyte through the porous carbon electrodes. Thus the overall reaction involves the combination of hydrogen gas and oxygen gas to form water. The cell runs continuously until the reactant's supply is exhausted. This type of cell operates efficiently in the temperature range 343 K to 413 K and provides a potential of about 0.9 V. AAEMFC is a type of AFC which employs a solid polymer electrolyte instead of aqueous potassium hydroxide (KOH) and it is superior to aqueous AFC.

High-temperature Fuel Cells

Solid Oxide Fuel Cell

Solid oxide fuel cells (SOFCs) use a solid material, most commonly a ceramic material called yttria-stabilized zirconia (YSZ), as the electrolyte. Because SOFCs are made entirely of solid materials, they are not limited to the flat plane configuration of other types of fuel cells and are often designed as rolled tubes. They require high operating temperatures (800–1000 °C) and can be run on a variety of fuels including natural gas.

SOFCs are unique since in those, negatively charged oxygen ions travel from the cathode (positive side of the fuel cell) to the anode (negative side of the fuel cell) instead of positively charged hydrogen ions travelling from the anode to the cathode, as is the case in all other types of fuel cells. Oxygen gas is fed through the cathode, where it absorbs electrons to create oxygen ions. The oxygen ions then travel through the electrolyte to react with hydrogen gas at the anode. The reaction at the anode produces electricity and water as by-products. Carbon dioxide may also be a by-product depending on the fuel, but the carbon emissions from an SOFC system are less than those from a fossil fuel combustion plant. The chemical reactions for the SOFC system can be expressed as follows:

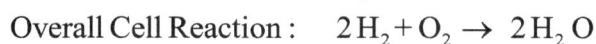

Anode Reaction : $\qquad 2H_2 + 2O^{2-} \rightarrow 2H_2O + 4e^-$

Cathode Reaction : $\qquad O_2 + 4e^- \rightarrow 2O^{2-}$

Overall Cell Reaction : $\quad 2H_2 + O_2 \rightarrow 2H_2O$

SOFC systems can run on fuels other than pure hydrogen gas. However, since hydrogen is necessary for the reactions listed above, the fuel selected must contain hydrogen atoms. For the fuel cell to operate, the fuel must be converted into pure hydrogen gas. SOFCs are capable of internally reforming light hydrocarbons such as methane (natural gas), propane and butane. These fuel cells are at an early stage of development.

Challenges exist in SOFC systems due to their high operating temperatures. One such challenge is the potential for carbon dust to build up on the anode, which slows down the internal reforming process. Research to address this "carbon coking" issue at the University of Pennsylvania has shown that the use of copper-based cermet (heat-resistant materials made of ceramic and metal) can reduce coking and the loss of performance. Another disadvantage of SOFC systems is slow start-up time, making SOFCs less useful for mobile applications. Despite these disadvantages, a high operating temperature provides an advantage by removing the need for a precious metal catalyst like platinum, thereby reducing cost. Additionally, waste heat from SOFC systems may be captured and reused, increasing the theoretical overall efficiency to as high as 80%–85%.

The high operating temperature is largely due to the physical properties of the YSZ electrolyte. As temperature decreases, so does the ionic conductivity of YSZ. Therefore, to obtain optimum performance of the fuel cell, a high operating temperature is required. According to their website, Ceres Power, a UK SOFC fuel cell manufacturer, has developed a method of reducing the operating temperature of their SOFC system to 500–600 degrees Celsius. They replaced the commonly used YSZ electrolyte with a CGO (cerium gadolinium oxide) electrolyte. The lower operating temperature allows them to use stainless steel instead of ceramic as the cell substrate, which reduces cost and start-up time of the system.

MCFC

Molten carbonate fuel cells (MCFCs) require a high operating temperature, 650 °C (1,200 °F), similar to SOFCs. MCFCs use lithium potassium carbonate salt as an electrolyte, and this salt liquefies at high temperatures, allowing for the movement of charge within the cell – in this case, negative carbonate ions.

Like SOFCs, MCFCs are capable of converting fossil fuel to a hydrogen-rich gas in the anode, eliminating the need to produce hydrogen externally. The reforming process creates CO2 emissions. MCFC-compatible fuels include natural gas, biogas and gas produced from coal. The hydrogen in the gas reacts with carbonate ions from the electrolyte to produce water, carbon dioxide, electrons and small amounts of other chemicals. The electrons travel through an external circuit creating electricity and return to the cathode. There, oxygen from the air and carbon dioxide recycled from the anode react with the electrons to form carbonate ions that replenish the electrolyte, completing the circuit. The chemical reactions for an MCFC system can be expressed as follows:

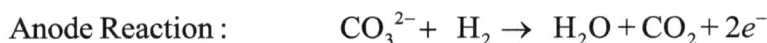

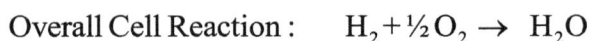

$$\text{Anode Reaction:} \quad CO_3^{2-} + H_2 \rightarrow H_2O + CO_2 + 2e^-$$

$$\text{Cathode Reaction:} \quad CO_2 + \tfrac{1}{2}O_2 + 2e^- \rightarrow CO_3^{2-}$$

$$\text{Overall Cell Reaction:} \quad H_2 + \tfrac{1}{2}O_2 \rightarrow H_2O$$

As with SOFCs, MCFC disadvantages include slow start-up times because of their high operating temperature. This makes MCFC systems not suitable for mobile applications, and this technology will most likely be used for stationary fuel cell purposes. The main challenge of MCFC technology is the cells' short life span. The high-temperature and carbonate electrolyte lead to corrosion of the anode and cathode. These factors accelerate the degradation of MCFC components, decreasing the durability and cell life. Researchers are addressing this problem by exploring corrosion-resistant materials for components as well as fuel cell designs that may increase cell life without decreasing performance.

MCFCs hold several advantages over other fuel cell technologies, including their resistance to impurities. They are not prone to "carbon coking", which refers to carbon build-up on the anode that results in reduced performance by slowing down the internal fuel reforming process. Therefore, carbon-rich fuels like gases made from coal are compatible with the system. The Department of Energy claims that coal, itself, might even be a fuel option in the future, assuming the system can be made resistant to impurities such as sulfur and particulates that result from converting coal into hydrogen. MCFCs also have relatively high efficiencies. They can reach a fuel-to-electricity efficiency of 50%, considerably higher than the 37–42% efficiency of a phosphoric acid fuel cell plant. Efficiencies can be as high as 65% when the fuel cell is paired with a turbine, and 85% if heat is captured and used in a Combined Heat and Power (CHP) system.

Fuel Cell Energy, a Connecticut-based fuel cell manufacturer, develops and sells MCFC fuel cells. The company says that their MCFC products range from 300 kW to 2.8 MW systems that achieve 47% electrical efficiency and can utilize CHP technology to obtain higher overall efficiencies. One product, the DFC-ERG, is combined with a gas turbine and, according to the company, it achieves an electrical efficiency of 65%.

Electric Storage Fuel Cell

The electric storage fuel cell is a conventional battery chargeable by electric power input, using the conventional electro-chemical effect. However, the battery further includes hydrogen (and oxygen) inputs for alternatively charging the battery chemically.

Efficiency of Leading Fuel Cell Types

Theoretical Maximum Efficiency

The energy efficiency of a system or device that converts energy is measured by the ratio of the amount of useful energy put out by the system ("output energy") to the total amount of energy that is put in ("input energy") or by useful output energy as a percentage of the total input energy. In the case of fuel cells, useful output energy is measured in electrical energy produced by the system. Input energy is the energy stored in the fuel. According to the U.S. Department of Energy, fuel cells are generally between 40–60% energy efficient. This is higher than some other systems for energy generation. For example, the typical internal combustion engine of a car is about 25% energy efficient. In combined heat and power (CHP) systems, the heat produced by the fuel cell is captured and put to use, increasing the efficiency of the system to up to 85–90%.

The theoretical maximum efficiency of any type of power generation system is never reached in practice, and it does not consider other steps in power generation, such as production, transportation and storage of fuel and conversion of the electricity into mechanical power. However, this calculation allows the comparison of different types of power generation. The maximum theoretical energy efficiency of a fuel cell is 83%, operating at low power density and using pure hydrogen and oxygen as reactants (assuming no heat recapture) According to the World Energy Council, this compares with a maximum theoretical efficiency of 58% for internal combustion engines.

In Practice

In a fuel-cell vehicle the tank-to-wheel efficiency is greater than 45% at low loads and shows average values of about 36% when a driving cycle like the NEDC (New European Driving Cycle) is used as test procedure. The comparable NEDC value for a Diesel vehicle is 22%. In 2008 Honda released a demonstration fuel cell electric vehicle (the Honda FCX Clarity) with fuel stack claiming a 60% tank-to-wheel efficiency.

It is also important to take losses due to fuel production, transportation, and storage into account. Fuel cell vehicles running on compressed hydrogen may have a power-plant-to-wheel efficiency of 22% if the hydrogen is stored as high-pressure gas, and 17% if it is stored as liquid hydrogen. Fuel cells cannot store energy like a battery, except as hydrogen, but in some applications, such as stand-alone power plants based on discontinuous sources such as solar or wind power, they are combined with electrolyzers and storage systems to form an energy storage system. Most hydrogen is used for oil refining, chemicals and fertilizer production and therefore produced by steam methane reforming, which emits carbon dioxide. The overall efficiency (electricity to hydrogen and back to electricity) of such plants (known as round-trip efficiency), using pure hydrogen and pure oxygen can be "from 35 up to 50 percent", depending on gas density and other conditions. The electrolyzer/fuel cell system can store indefinite quantities of hydrogen, and is therefore suited for long-term storage.

Solid-oxide fuel cells produce heat from the recombination of the oxygen and hydrogen. The ceramic can run as hot as 800 degrees Celsius. This heat can be captured and used to heat water in a micro combined heat and power (m-CHP) application. When the heat is captured, total efficiency can reach 80–90% at the unit, but does not consider production and distribution losses. CHP units are being developed today for the European home market.

Professor Jeremy P. Meyers, in the Electrochemical Society journal Interface in 2008, wrote, "While fuel cells are efficient relative to combustion engines, they are not as efficient as batteries, due primarily to the inefficiency of the oxygen reduction reaction (and the oxygen evolution reaction, should the hydrogen be formed by electrolysis of water). They make the most sense for operation disconnected from the grid, or when fuel can be provided continuously. For applications that require frequent and relatively rapid start-ups, where zero emissions are a requirement, as in enclosed spaces such as warehouses, and where hydrogen is considered an acceptable reactant, a (PEM fuel cell) is becoming an increasingly attractive choice (if exchanging batteries is inconvenient)". In 2013 military organizations were evaluating fuel cells to determine if they could significantly reduce the battery weight carried by soldiers.

Applications

Power

Type 212 submarine with fuel cell propulsion of the German Navy in dry dock.

Stationary fuel cells are used for commercial, industrial and residential primary and backup power generation. Fuel cells are very useful as power sources in remote locations, such as spacecraft, remote weather stations, large parks, communications centers, rural locations including research stations, and in certain military applications. A fuel cell system running on hydrogen can be compact and lightweight, and have no major moving parts. Because fuel cells have no moving parts and do not involve combustion, in ideal conditions they can achieve up to 99.9999% reliability. This equates to less than one minute of downtime in a six-year period.

Since fuel cell electrolyzer systems do not store fuel in themselves, but rather rely on external storage units, they can be successfully applied in large-scale energy storage, rural areas being one example. There are many different types of stationary fuel cells so efficiencies vary, but most are between 40% and 60% energy efficient. However, when the fuel cell's waste heat is used to heat a building in a cogeneration system this efficiency can increase to 85%. This is significantly more efficient than traditional coal power plants, which are only about one third energy efficient. Assuming production at scale, fuel cells could save 20–40% on energy costs when used in cogeneration systems. Fuel cells are also much cleaner than traditional power generation; a fuel cell power plant using natural gas as a hydrogen source would create less than one ounce of pollution (other than CO_2) for every 1,000 kW·h produced, compared to 25 pounds of pollutants generated by conventional combustion systems. Fuel Cells also produce 97% less nitrogen oxide emissions than conventional coal-fired power plants.

One such pilot program is operating on Stuart Island in Washington State. There the Stuart Island Energy Initiative has built a complete, closed-loop system: Solar panels power an electrolyzer, which makes hydrogen. The hydrogen is stored in a 500-U.S.-gallon (1,900 L) tank at 200 pounds per square inch (1,400 kPa), and runs a ReliOn fuel cell to provide full electric back-up to the off-the-grid residence. Another closed system loop was unveiled in late 2011 in Hempstead, NY.

Fuel cells can be used with low-quality gas from landfills or waste-water treatment plants to generate power and lower methane emissions. A 2.8 MW fuel cell plant in California is said to be the largest of the type.

Cogeneration

Combined heat and power (CHP) fuel cell systems, including Micro combined heat and power (MicroCHP) systems are used to generate both electricity and heat for homes, office building and factories. The system generates constant electric power (selling excess power back to the grid when it is not consumed), and at the same time produces hot air and water from the waste heat. As the result CHP systems have the potential to save primary energy as they can make use of waste heat which is generally rejected by thermal energy conversion systems. A typical capacity range of home fuel cell is 1–3 kW_{el}/4–8 kW_{th}. CHP systems linked to absorption chillers use their waste heat for refrigeration.

The waste heat from fuel cells can be diverted during the summer directly into the ground providing further cooling while the waste heat during winter can be pumped directly into the building. The University of Minnesota owns the patent rights to this type of system.

Co-generation systems can reach 85% efficiency (40–60% electric + remainder as thermal). Phosphoric-acid fuel cells (PAFC) comprise the largest segment of existing CHP products worldwide and can provide combined efficiencies close to 90%. Molten Carbonate (MCFC) and Solid Oxide Fuel Cells (SOFC) are also used for combined heat and power generation and have electrical energy efficiencies around 60%. Disadvantages of co-generation systems include slow ramping up and down rates, high cost and short lifetime. Also their need to have a hot water storage tank to smooth out the thermal heat production was a serious disadvantage in the domestic market place where space in domestic properties is at a great premium.

Delta-ee consultants stated in 2013 that with 64% of global sales the fuel cell micro-combined heat and power passed the conventional systems in sales in 2012. The Japanese ENE FARM project will pass 100,000 FC mCHP systems in 2014, 34.213 PEMFC and 2.224 SOFC were installed in the period 2012-2014, 30,000 units on LNG and 6,000 on LPG.

Automobiles

As of 2017, about 6500 FCEVs have been leased or sold worldwide. Three fuel cell electric vehicles have been introduced for commercial lease and sale: the Honda Clarity, Toyota Mirai and the Hyundai ix35 FCEV. Additional demonstration models include the Honda FCX Clarity, and Mercedes-Benz F-Cell. As of June 2011 demonstration FCEVs had driven more than 4,800,000 km (3,000,000 mi), with more than 27,000 refuelings. Fuel cell electric vehicles feature an average range of 314 miles between refuelings. They can be refueled in less than 5 minutes. The U.S. Department of Energy's Fuel Cell Technology Program states that, as of 2011, fuel cells achieved 53–59% efficiency at one-quarter power and 42–53% vehicle efficiency at full power, and a durability of over 120,000 km (75,000 mi) with less than 10% degradation. In a Well-to-Wheels simulation analysis that "did not address the economics and market constraints", General Motors and its partners estimated that per mile traveled, a fuel cell electric vehicle running on compressed gaseous hydrogen produced from natural gas could use about 40% less energy and emit 45% less greenhouse gasses than an internal combustion vehicle. A lead engineer from the Department of Energy whose team is testing fuel cell cars said in 2011 that the potential appeal is that "these are full-function vehicles with no limitations on range or refueling rate so they are a direct replacement for any vehicle. For instance, if you drive a full sized SUV and pull a boat up into the mountains, you can do that with this technology and you can't with current battery-only vehicles, which are more geared toward city driving".

Element One fuel cell vehicle.

In 2015, Toyota introduced its first fuel cell vehicle, the Mirai, at a price of $57,000. Hyundai introduced the limited production Hyundai ix35 FCEV under a lease agreement. In 2016, Honda started leasing the Honda Clarity Fuel Cell.

Some commentators believe that hydrogen fuel cell cars will never become economically competitive with other technologies or that it will take decades for them to become profitable. Elon Musk, CEO of battery-electric vehicle maker Tesla Motors, stated in 2015 that fuel cells for use in cars will never be commercially viable because of the inefficiency of producing, transporting and storing hydrogen and the flammability of the gas, among other reasons. Professor Jeremy P. Meyers estimated in 2008 that cost reductions over a production ramp-up period will take about 20 years after fuel-cell cars are introduced before they will be able to compete commercially with current market technologies, including gasoline internal combustion engines. In 2011, the chairman and CEO of General Motors, Daniel Akerson, stated that while the cost of hydrogen fuel cell cars is decreasing: "The car is still too expensive and probably won't be practical until the 2020-plus period, I don't know".

In 2012, Lux Research, Inc. issued a report that stated: "The dream of a hydrogen economy is no nearer". It concluded that "Capital cost will limit adoption to a mere 5.9 GW" by 2030, providing "a nearly insurmountable barrier to adoption, except in niche applications". The analysis concluded that, by 2030, PEM stationary market will reach $1 billion, while the vehicle market, including forklifts, will reach a total of $2 billion. Other analyses cite the lack of an extensive hydrogen infrastructure in the U.S. as an ongoing challenge to Fuel Cell Electric Vehicle commercialization. In 2006, a study for the IEEE showed that for hydrogen produced via electrolysis of water: "Only about 25% of the power generated from wind, water, or sun is converted to practical use". The study further noted that "Electricity obtained from hydrogen fuel cells appears to be four times as expensive as electricity drawn from the electrical transmission grid. Because of the high energy losses [hydrogen] cannot compete with electricity". Furthermore, the study found: "Natural gas reforming is not a sustainable solution". "The large amount of energy required to isolate hydrogen from natural compounds (water, natural gas, biomass), package the light gas by compression or liquefaction, transfer the energy carrier to the user, plus the energy lost when it is converted to useful electricity with fuel cells, leaves around 25% for practical use".

In 2014, Joseph Romm, the author of The Hype About Hydrogen, stated that FCVs still had not overcome the high fueling cost, lack of fuel-delivery infrastructure, and pollution caused by producing hydrogen. "It would take several miracles to overcome all of those problems simultaneously in the coming decades". He concluded that renewable energy cannot economically be used to make hydrogen for an FCV fleet "either now or in the future". Greentech Media's analyst reached similar conclusions in 2014. In 2015, Clean Technica listed some of the disadvantages of hydrogen fuel cell vehicles. So did Car Throttle.

Buses

As of August 2011, there were a total of approximately 100 fuel cell buses deployed around the world. Most buses are produced by UTC Power, Toyota, Ballard, Hydrogenics, and Proton Motor. UTC buses had accumulated over 970,000 km (600,000 mi) of driving by 2011. Fuel cell buses have a 39–141% higher fuel economy than diesel buses and natural gas buses. Fuel cell buses have been deployed around the world including in Whistler, Canada; San Francisco, United States; Hamburg, Germany; Shanghai, China; London, England; and São Paulo, Brazil.

The Fuel Cell Bus Club is a global cooperative effort in trial fuel cell buses. Notable projects include:

- 12 fuel cell buses are being deployed in the Oakland and San Francisco Bay area of California.

- In 2007, Daimler AG, with 36 experimental buses powered by Ballard Power Systems fuel cells in eleven cities, completed a successful three-year trial.

- A fleet of Thor buses with UTC Power fuel cells was deployed in California, operated by SunLine Transit Agency.

Toyota FCHV-BUS at the Expo 2005.

Forklifts

A fuel cell forklift (also called a fuel cell lift truck) is a fuel cell-powered industrial forklift truck used to lift and transport materials. In 2013 there were over 4,000 fuel cell forklifts used in material handling in the US, of which only 500 received funding from DOE. The global market is 1 million fork lifts per year. Fuel cell fleets are operated by various companies, including Sysco Foods, FedEx Freight, GENCO (at Wegmans, Coca-Cola, Kimberly Clark, and Whole Foods), and H-E-B Grocers. Europe demonstrated 30 fuel cell forklifts with Hylift and extended it with HyLIFT-EUROPE to 200 units, with other projects in France and Austria. Pike Research stated in 2011 that fuel cell-powered forklifts will be the largest driver of hydrogen fuel demand by 2020.

Most companies in Europe and the US do not use petroleum-powered forklifts, as these vehicles work indoors where emissions must be controlled and instead use electric forklifts. Fuel cell-powered forklifts can provide benefits over battery-powered forklifts as they can work for a full 8-hour shift on a single tank of hydrogen and can be refueled in 3 minutes. Fuel cell-powered forklifts can be used in refrigerated warehouses, as their performance is not degraded by lower temperatures. The FC units are often designed as drop-in replacements.

Motorcycles and Bicycles

In 2005 a British manufacturer of hydrogen-powered fuel cells, Intelligent Energy (IE), produced the first working hydrogen-run motorcycle called the ENV (Emission Neutral Vehicle). The motorcycle holds enough fuel to run for four hours, and to travel 160 km (100 mi) in an urban area, at a top speed of 80 km/h (50 mph). In 2004 Honda developed a fuel-cell motorcycle that utilized the Honda FC Stack.

Other examples of motorbikes and bicycles that use hydrogen fuel cells include the Taiwanese company APFCT's scooter using the fueling system from Italy's Acta SpA and the Suzuki Burgman scooter with an IE fuel cell that received EU Whole Vehicle Type Approval in 2011. Suzuki Motor Corp. and IE have announced a joint venture to accelerate the commercialization of zero-emission vehicles.

Airplanes

In 2003, the world's first propeller-driven airplane to be powered entirely by a fuel cell was flown. The fuel cell was a stack design that allowed the fuel cell to be integrated with the plane's aerodynamic surfaces. Fuel cell-powered unmanned aerial vehicles (UAV) include a Horizon fuel cell UAV that set the record distance flown for a small UAV in 2007. Boeing researchers and industry partners throughout Europe conducted experimental flight tests in February 2008 of a manned airplane powered only by a fuel cell and lightweight batteries. The fuel cell demonstrator airplane, as it was called, used a proton exchange membrane (PEM) fuel cell/lithium-ion battery hybrid system to power an electric motor, which was coupled to a conventional propeller.

In 2009 the Naval Research Laboratory's (NRL's) Ion Tiger utilized a hydrogen-powered fuel cell and flew for 23 hours and 17 minutes. Fuel cells are also being tested and considered to provide auxiliary power in aircraft, replacing fossil fuel generators that were previously used to start the engines and power on board electrical needs, while reducing carbon emissions.

In 2016 a Raptor E1 drone made a successful test flight using a fuel cell that was lighter than the lithium-ion battery it replaced. The flight lasted 10 minutes at an altitude of 80 metres (260 ft), although the fuel cell reportedly had enough fuel to fly for two hours. The fuel was contained in approximately 100 solid 1 square centimetre (0.16 sq in) pellets composed of a proprietary chemical within an unpressurized cartridge. The pellets are physically robust and operate at temperatures as warm as 50 °C (122 °F). The cell was from Arcola Energy.

Boats

The world's first certified fuel cell boat (HYDRA), in Leipzig/Germany.

The world's first fuel-cell boat HYDRA used an AFC system with 6.5 kW net output. Iceland has committed to converting its vast fishing fleet to use fuel cells to provide auxiliary power by 2015 and, eventually, to provide primary power in its boats. Amsterdam recently introduced its first fuel cell-powered boat that ferries people around the city's canals.

Submarines

The Type 212 submarines of the German and Italian navies use fuel cells to remain submerged for weeks without the need to surface.

The U212A is a non-nuclear submarine developed by German naval shipyard Howaldtswerke Deutsche Werft. The system consists of nine PEM fuel cells, providing between 30 kW and 50 kW each. The ship is silent, giving it an advantage in the detection of other submarines. A naval paper has theorized about the possibility of a nuclear-fuel cell hybrid whereby the fuel cell is used when silent operations are required and then replenished from the Nuclear reactor (and water).

Portable Power Systems

Portable fuel cell systems are generally classified as weighing under 10 kg and providing power of less than 5 kW. The potential market size for smaller fuel cells is quite large with an up to 40% per annum potential growth rate and a market size of around $10 billion, leading a great deal of research to be devoted to the development of portable power cells. Within this market two groups have been identified. The first is the microfuel cell market, in the 1-50 W range for power smaller electronic devices. The second is the 1-5 kW range of generators for larger scale power generation (e.g. military outposts, remote oil fields).

Microfuel cells are primarily aimed at penetrating the market for phones and laptops. This can be primarily attributed to the advantageous energy density provided by fuel cells over a lithium-ion battery, for the entire system. For a battery, this system includes the charger as well as the battery itself. For the fuel cell this system would include the cell, the necessary fuel and peripheral attachments. Taking the full system into consideration, fuel cells have been shown to provide 530Wh/kg compared to 44 Wh/kg for lithium ion batteries. However, while the weight of fuel cell systems offer a distinct advantage the current costs are not in their favor. while a battery system will generally cost around $1.20 per Wh, fuel cell systems cost around $5 per Wh, putting them at a significant disadvantage.

As power demands for cell phones increase, fuel cells could become much more attractive options for larger power generation. The demand for longer on time on phones and computers is something often demanded by consumers so fuel cells could start to make strides into laptop and cell phone markets. The price will continue to go down as developments in fuel cells continues to accelerate. Current strategies for improving micro fuelcells is through the use of carbon nanotubes. It was shown by Girishkumar et al. that depositing nanotubes on electrode surfaces allows for substantially greater surface area increasing the oxygen reduction rate.

Fuel cells for use in larger scale operations also show much promise. Portable power systems that use fuel cells can be used in the leisure sector (i.e. RVs, cabins, marine), the industrial sector (i.e. power for remote locations including gas/oil wellsites, communication towers, security, weather stations), and in the military sector. SFC Energy is a German manufacturer of direct methanol fuel cells for a variety of portable power systems. Ensol Systems Inc. is an integrator of portable power systems, using the SFC Energy DMFC. The key advantage of fuel cells in this market is the great power generation per weight. While fuel cells can be expensive, for remote locations that require dependable energy fuel cells hold great power. For a 72-h excursion the comparison in weight is substantial, with a fuel cell only weighing 15 pounds compared to 29 pounds of batteries needed for the same energy.

Other Applications

- Providing power for base stations or cell sites.

- Distributed generation.

- Emergency power systems are a type of fuel cell system, which may include lighting, generators and other apparatus, to provide backup resources in a crisis or when regular systems fail. They find uses in a wide variety of settings from residential homes to hospitals, scientific laboratories, data centers.

- Telecommunication equipment and modern naval ships.

- An uninterrupted power supply (UPS) provides emergency power and, depending on the topology, provide line regulation as well to connected equipment by supplying power from a separate source when utility power is not available. Unlike a standby generator, it can provide instant protection from a momentary power interruption.

- Base load power plants.

- Solar Hydrogen Fuel Cell Water Heating.

- Hybrid vehicles, pairing the fuel cell with either an ICE or a battery.

- Notebook computers for applications where AC charging may not be readily available.

- Portable charging docks for small electronics (e.g. a belt clip that charges a cell phone or PDA).

- Smartphones, laptops and tablets.

- Small heating appliances.

- Food preservation, achieved by exhausting the oxygen and automatically maintaining oxygen exhaustion in a shipping container, containing, for example, fresh fish.

- Breathalyzers, where the amount of voltage generated by a fuel cell is used to determine the concentration of fuel (alcohol) in the sample.

- Carbon monoxide detector, electrochemical sensor.

Fueling Stations

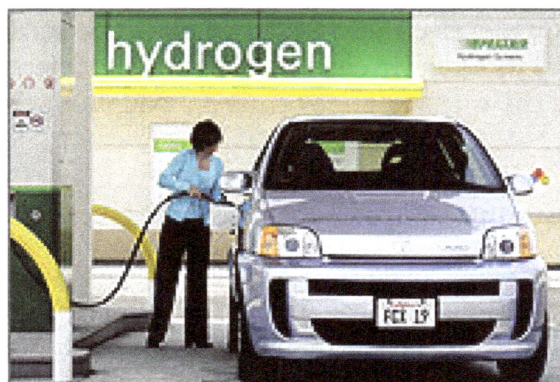
Hydrogen fueling station.

In 2013, The New York Times reported that there were "10 hydrogen stations available to the public in the entire United States: one in Columbia, S.C., eight in Southern California and the one in

Emeryville". As of December 2016, there were 31 publicly accessible hydrogen refueling stations in the US, 28 of which were located in California.

A public hydrogen refueling station in Iceland operated from 2003 to 2007. It served three buses in the public transport net of Reykjavík. The station produced its own hydrogen with an electrolyzing unit. The 14 stations in Germany were planned to be expanded to 50 by 2015 through its public–private partnership Now GMBH.

By May 2017, there were 91 hydrogen fueling stations in Japan. As of 2016, Norway planned to build a network of hydrogen stations between the major cities, starting in 2017.

ELECTROCHEMICAL GAS SENSOR

Electrochemical gas sensors are gas detectors that measure the concentration of a target gas by oxidizing or reducing the target gas at an electrode and measuring the resulting current.

Construction

The sensors contain two or three electrodes, occasionally four, in contact with an electrolyte. The electrodes are typically fabricated by fixing a high surface area precious metal on to the porous hydrophobic membrane. The working electrode contacts both the electrolyte and the ambient air to be monitored usually via a porous membrane. The electrolyte most commonly used is a mineral acid, but organic electrolytes are also used for some sensors. The electrodes and housing are usually in a plastic housing which contains a gas entry hole for the gas and electrical contacts.

Theory of Operation

The gas diffuses into the sensor, through the back of the porous membrane to the working electrode where it is oxidized or reduced. This electrochemical reaction results in an electric current that passes through the external circuit. In addition to measuring, amplifying and performing other signal processing functions, the external circuit maintains the voltage across the sensor between the working and counter electrodes for a two electrode sensor or between the working and reference electrodes for a three electrode cell. At the counter electrode an equal and opposite reaction occurs, such that if the working electrode is an oxidation, then the counter electrode is a reduction.

Diffusion Controlled Response

The magnitude of the current is controlled by how much of the target gas is oxidized at the working electrode. Sensors are usually designed so that the gas supply is limited by diffusion and thus the output from the sensor is linearly proportional to the gas concentration. This linear output is one of the advantages of electrochemical sensors over other sensor technologies, (e.g. infrared), whose output must be linearized before they can be used. A linear output allows for more precise measurement of low concentrations and much simpler calibration (only baseline and one point are needed).

Diffusion control offers another advantage. Changing the diffusion barrier allows the sensor manufacturer to tailor the sensor to a particular target gas concentration range. In addition, since the diffusion barrier is primarily mechanical, the calibration of electrochemical sensors tends to be more stable over time and so electrochemical sensor based instruments require much less maintenance than some other detection technologies. In principle, the sensitivity can be calculated based on the diffusion properties of the gas path into the sensor, though experimental errors in the measurement of the diffusion properties make the calculation less accurate than calibrating with test gas.

Cross Sensitivity

For some gases such as ethylene oxide, cross sensitivity can be a problem because ethylene oxide requires a very active working electrode catalyst and high operating potential for its oxidation. Therefore gases which are more easily oxidized such as alcohols and carbon monoxide will also give a response. Cross sensitivity problems can be eliminated though through the use of a chemical filter, for example filters that allows the target gas to pass through unimpeded, but which reacts with and removes common interferences.

While electrochemical sensors offer many advantages, they are not suitable for every gas. Since the detection mechanism involves the oxidation or reduction of the gas, electrochemical sensors are usually only suitable for gases which are electrochemically active, though it is possible to detect electrochemically inert gases indirectly if the gas interacts with another species in the sensor that then produces a response. Sensors for carbon dioxide are an example of this approach and they have been commercially available for several years.

FLOW BATTERY

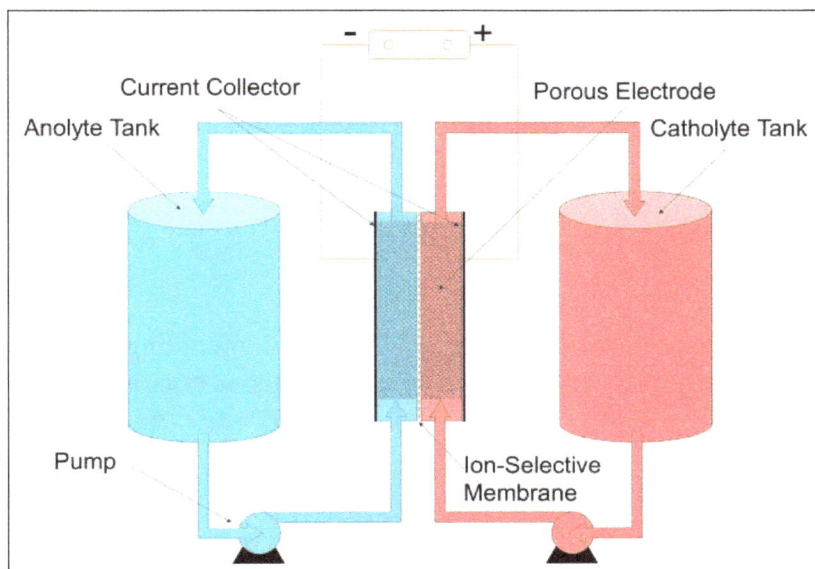

A typical flow battery consists of two tanks of liquids which are
pumped past a membrane held between two electrodes.

A flow battery, or redox flow battery (after reduction–oxidation), is a type of electrochemical cell where chemical energy is provided by two chemical components dissolved in liquids contained within the system and separated by a membrane. Ion exchange (accompanied by flow of electric current) occurs through the membrane while both liquids circulate in their own respective space. Cell voltage is chemically determined by the Nernst equation and ranges, in practical applications, from 1.0 to 2.2 volts.

A flow battery may be used like a fuel cell (where the spent fuel is extracted and new fuel is added to the system) or like a rechargeable battery (where an electric power source drives regeneration of the fuel). While it has technical advantages over conventional rechargeables, such as potentially separable liquid tanks and near unlimited longevity, current implementations are comparatively less powerful and require more sophisticated electronics.

The energy capacity is a function of the electrolyte volume (amount of liquid electrolyte), and the power is a function of the surface area of the electrodes.

Construction Principle

A flow battery is a rechargeable fuel cell in which an electrolyte containing one or more dissolved electroactive elements flows through an electrochemical cell that reversibly converts chemical energy directly to electricity (electroactive elements are "elements in solution that can take part in an electrode reaction or that can be adsorbed on the electrode"). Additional electrolyte is stored externally, generally in tanks, and is usually pumped through the cell (or cells) of the reactor, although gravity feed systems are also known. Flow batteries can be rapidly "recharged" by replacing the electrolyte liquid (in a similar way to refilling fuel tanks for internal combustion engines) while simultaneously recovering the spent material for re-energization. Many flow batteries use carbon felt electrodes due to its low cost and adequate electrical conductivity, although these electrodes somewhat limit the charge/discharge power due to their low inherent activity towards many redox couples.

In other words, a flow battery is just like an electrochemical cell, with the exception that the ionic solution (electrolyte) is not stored in the cell around the electrodes. Rather, the ionic solution is stored outside of the cell, and can be fed into the cell in order to generate electricity. The total amount of electricity that can be generated depends on the size of the storage tanks.

Flow batteries are governed by the design principles established by electrochemical engineering.

Types

Various types of flow cells (batteries) have been developed, including redox, hybrid and membraneless. The fundamental difference between conventional batteries and flow cells is that energy is stored in the electrode material in conventional batteries, while in flow cells it is stored in the electrolyte.

Redox

The redox (reduction–oxidation) cell is a reversible cell in which electrochemical components are dissolved in the electrolyte. Redox flow batteries are rechargeable (secondary cells). Because they employ heterogeneous electron transfer rather than solid-state diffusion or intercalation they are more appropriately called fuel cells rather than batteries. The performance of a solid-state battery depends on the diffusion of ions within the electrolyte, solid electrolytes must have high ionic conductivity, very low electronic conductivity and a high degree of chemical stability. In industrial practice, fuel cells are usually, and unnecessarily, considered to be primary cells, such as the H_2/O_2 system. The unitized regenerative fuel cell on NASA's Helios Prototype is another reversible fuel cell. The European Patent Organisation classifies redox flow cells (H01M8/18C4) as a sub-class of regenerative fuel cells (H01M8/18). Examples of redox flow batteries are the Vanadium redox flow battery, polysulfide bromide battery (Regenesys), and uranium redox flow battery. Redox fuel cells are less common commercially although many systems have been proposed.

Vanadium Redox Flow Batteries are the most marketed flow batteries at present, due to a number of advantages they present on other chemistries, despite their limited energy and power densities. Since they use vanadium at both electrodes, they do not suffer cross-contamination issues. For the same reason, they have unparalleled cycle lives (15,000–20,000 cycles), which in turns results in record levelized costs of energy (LCOE, i.e. the system cost divided by the usable energy, the cycle life, and round-trip efficiency), which are in the order of a few tens of \$ cents or € cents per kWh, namely much lower than other batteries solid-state batteries and not so far from the targets of \$0.05 and €0.05, stated by US and EC government agencies.

A prototype zinc-polyiodide flow battery has been demonstrated with an energy density of 167 Wh/l (watt-hours per liter). Older zinc-bromide cells reach 70 Wh/l. For comparison, lithium iron phosphate batteries store 233 Wh/l. The zinc-polyiodide battery is claimed to be safer than other flow batteries given its absence of acidic electrolytes, nonflammability and operating range of −4 to 122 °F (−20 to 50 °C) that does not require extensive cooling circuitry, which would add weight and occupy space. One unresolved issue is zinc build-up on the negative electrode that permeated the membrane, reducing efficiency. Because of the Zn dendrite formation, the Zn-halide batteries cannot operate at high current density (> 20 mA/cm²) and thus have limited power density. Adding alcohol to the electrolyte of the ZnI battery can slightly control the problem.

When the battery is fully discharged, both tanks hold the same electrolyte solution: a mixture of positively charged zinc ions (Zn^{2+}) and negatively charged iodide ion, I-. When charged, one tank holds another negative ion, polyiodide, I3-. The battery produces power by pumping liquid from external tanks into the battery's stack area where the liquids are mixed. Inside the stack, zinc ions pass through a selective membrane and change into metallic zinc on the stack's negative side. To further increase the energy density of the zinc-iodide flow battery, bromide ions (Br−) are used as the complexing agent to stabilize the free iodine, forming iodine-bromide ions (I2Br−) as a means to free up iodide ions for charge storage.

Traditional flow battery chemistries have both low specific energy (which makes them too heavy for fully electric vehicles) and low specific power (which makes them too expensive for stationary energy storage). However a high power of 1.4 W/cm2 was demonstrated for hydrogen-bromine flow batteries, and a high specific energy (530 Wh/kg at the tank level) was shown for hydrogen-bromate flow batteries

One system uses organic polymers and a saline solution with a cellulose membrane. The prototype withstood 10000 charging cycles while retaining substantial capacity. The energy density was 10 Wh/l. Current density reached 100 milliamperes/cm2.

Hybrid

The hybrid flow battery uses one or more electroactive components deposited as a solid layer. In this case, the electrochemical cell contains one battery electrode and one fuel cell electrode. This type is limited in energy by the surface area of the electrode. Hybrid flow batteries include the zinc-bromine, zinc–cerium, lead–acid, and iron-salt flow batteries. Weng et al. reported a Vanadium-Metal hydride rechargeable hybrid flow battery with an experimental OCV of 1.93 V and operating voltage of 1.70 V, very high values among rechargeable flow batteries with aqueous electrolytes. This hybrid battery consists of a graphite felt positive electrode operating in a mixed solution of $VOSO_4$ and H_2SO_4, and a metal hydride negative electrode in KOH aqueous solution. The two electrolytes of different pH are separated by a bipolar membrane. The system demonstrated good reversibility and high efficiencies in coulomb (95%), energy (84%), and voltage (88%). They reported further improvements of this new redox couple with achievements of increased current density, operation of larger 100 cm² electrodes, and the operation of 10 large cells in series. Preliminary data using a fluctuating simulated power input tested the viability toward kWh scale storage. Recently, a high energy density Mn(VI)/Mn(VII)-Zn hybrid flow battery has been proposed.

Membraneless

A membraneless battery relies on laminar flow in which two liquids are pumped through a channel, where they undergo electrochemical reactions to store or release energy. The solutions stream through in parallel, with little mixing. The flow naturally separates the liquids, eliminating the need for a membrane.

Membranes are often the most costly and least reliable components of batteries, as they can be corroded by repeated exposure to certain reactants. The absence of a membrane enables the use of a liquid bromine solution and hydrogen: this combination is problematic when membranes are used, because they form hydrobromic acid that can destroy the membrane. Both materials are available at low cost.

The design uses a small channel between two electrodes. Liquid bromine flows through the channel over a graphite cathode and hydrobromic acid flows under a porous anode. At the same time, hydrogen gas flows across the anode. The chemical reaction can be reversed to recharge the battery — a first for any membraneless design. One such membraneless flow battery published in August 2013 produced a maximum power density of 7950 W/m², three times as much power as other membraneless systems— and an order of magnitude higher than lithium-ion batteries.

Recently, a macroscale membraneless redox flow battery capable of recharging and recirculation of the same electrolyte streams for multiple cycles has been demonstrated. The battery is based on immiscible organic catholyte and aqueous anolyte liquids, which exhibits high capacity retention and Coulombic efficiency during cycling.

Primus Power has developed patented technology in its zinc bromine flow battery, a type of redox flow battery, to eliminate the membrane or separator, which reduces costs and failure rates. The Primus Power membraneless redox flow battery is working in installations in the United States and Asia with a second generation product announced 21 February 2017.

Organic

Compared to traditional aqueous inorganic redox flow batteries such as vanadium redox flow batteries and Zn-Br2 batteries, which have been developed for decades, organic redox flow batteries emerged in 2009 and hold great promise to overcome major drawbacks preventing economical and extensive deployment of traditional inorganic redox flow batteries. The primary merit of organic redox flow batteries lies in the tunable redox properties of the redox-active components.

Organic redox flow batteries could be further classified into two categories: Aqueous Organic Redox Flow Batteries (AORFBs) and Non-aqueous Organic Redox Flow Batteries (NAORFBs). AORFBs use water as solvent for electrolyte materials while NAORFBs employ organic solvents to dissolve redox active materials. Depending on using one or two organic redox active electrolytes as anode and cathode, AORFBs and NAORFBs can be further divided into total organic systems and hybrid organic systems that use inorganic materials for anode or cathode. In larger-scale energy storage, Due to lower solvent cost and higher conductivity AORFBs have greater commercial potential than NAORFBs , as well as the safety advantages of water-based electrolytes over non-aqueous electrolytes. The advantage of NAORFBs lies in their much larger voltage window and ability to possibly occupy less physical space for an installed storage facility. The content below materials advances for these organic based systems.

Quinones and their derivatives are the basis of many organic redox systems including NARFBs and AORFBs. In one study, 1, 2-dihydrobenzoquinone-3, 5-disulfonic acid (BQDS) and 1, 4-dihydrobenzoquinone-2-sulfonic acid (BQS) were employed as cathodes, and conventional $Pb/PbSO_4$ was the anolyte in an acid AORFB. These first AORFBs are hybrid systems as they use organic redox active materials only for the cathode side. The quinones accepts two units of electrical charge, compared with one in conventional catholyte, implying that such a battery could store twice as much energy in a given volume.

9,10-Anthraquinone-2,7-disulfonic acid (AQDS), also a quinone, has been evaluated as well. AQDS undergoes rapid, reversible two-electron/two-proton reduction on a glassy carbon electrode in sulfuric acid. An aqueous flow battery with inexpensive carbon electrodes, combining the quinone/hydroquinone couple with the Br_2/Br^{-redox} couple, yields a peak galvanic power density exceeding 6,000 W/m^2 at 13,000 A/m^2. Cycling showed >99 % storage capacity retention per cycle. Volumetric energy density was over 20 Wh/l. Anthraquinone-2-sulfonic acid and anthraquinone-2,6-disulfonic acid on the negative side and 1,2-dihydrobenzoquinone- 3,5-disulfonic acid on the positive side avoids the use of hazardous Br_2. The battery was claimed to last for 1,000 cycles without degradation although no official data were published. While this total organic system appears robust, it has a low cell voltage (ca. 0.55 V) and a low energy density (< 4 Wh/L).

Hydrobromic acid used as an electrolyte has been replaced with a far less toxic alkaline solution (1M KOH) and ferrocyanide. The higher pH is less corrosive, allowing the use of inexpensive polymer tanks. The increased electrical resistance in the membrane was compensated by increasing the voltage. The cell voltage was 1.2 V. The cell's efficiency exceeded 99%, while round-trip efficiency measured 84%. The battery has an expected lifetime of at least 1,000 cycles. Its theoretic energy density was 19 Wh per liter. Ferrocyanide's chemically stability in high pH KOH solution without forming $Fe(OH)_2$ or $Fe(OH)_3$ needs to be verified before scale-up.

Another organic AORFB has been demonstrated methyl viologen as anolyte and 4-hydroxy-2, 2, 6, 6-tetramethylpiperidin-1-oxyl as catholyte, plus sodium chloride and a low-cost anion exchange membrane to enable charging and discharging. This MV/TEMPO system has the highest cell voltage, 1.25 V, and, possibly, lowest capital cost ($180/kWh) reported for AORFBs. The water-based liquid electrolytes were designed as a drop-in replacement for current systems without replacing existing infrastructure. A 600-milliwatt test battery was stable for 100 cycles with nearly 100 percent efficiency at current densities ranging from 20 to 100 mA per square centimeter, with optimal performance rated at 40-50 mA, at which about 70 percent of the battery's original voltage was retained. The significance of the research is that neutral AORFBs can be more environmentally friendly than acid or alkaline AORFBs while showing electrochemical performance comparable to corrosive acidic or alkaline RFBs. The MV/TEMPO AORFB has an energy density of 8.4 Wh/L with the limitation on the TEMPO side. The next step is to identify a high capacity catholyte to match MV (ca. 3.5 M solubility in water, 93.8 Ah/L).

One flow-battery concept is based on redox active, organic polymers employs viologen and TEMPO with dialysis membranes. The polymer-based redox-flow battery (pRFB) uses functionalized macromolecules (similar to acrylic glass or Styrofoam) being dissolved in water as active material for the anode as well as the cathode. Thereby, metals and strongly corrosive electrolytes – like vanadium salts in sulfuric acid – are avoided and simple dialysis membranes can be employed. The membrane, which separates the cathode and the anode of the flow cell, works like a strainer and is produced much more easily and at lower cost than conventional ion-selective membranes. It retains the big "spaghetti"-like polymer molecules, while allowing the small counterions to pass. The concept may solve the high cost of traditional Nafion membrane but the design and synthesis of redox active polymer with high solubility in water is not trivial.

Aligned with the tunability of the redox-active components as the main advantage of organic redox flow batteries, the idea of integrating both anolyte and catholyte in the same molecule has been developed. Those so called, bifunctional analytes or combi-molecules allow to use the same material in both tanks, which definitely has relevant advantages on the battery performance, as diminishing the effect of crossover. Thus, diaminoanthraquinone, also a quinone, and indigo based molecules as well as TEMPO/phenazine combining molecules have been presented as potential electrolytes for the development of symmetric redox-flow batteries (SRFB).

Metal Hydride

Proton flow batteries (PFB) integrate a metal hydride storage electrode into a reversible proton exchange membrane (PEM) fuel cell. During charging, PFB combines hydrogen ions produced from splitting water with electrons and metal particles in one electrode of a fuel cell. The energy is stored in the form of a solid-state metal hydride. Discharge produces electricity and water when the process is reversed and the protons are combined with ambient oxygen. Metals less expensive than lithium can be used and provide greater energy density than lithium cells.

Nano-network

Lithium–sulfur system arranged in a network of nanoparticles eliminates the requirement that charge moves in and out of particles that are in direct contact with a conducting plate. Instead, the nanoparticle network allows electricity to flow throughout the liquid. This allows more energy to be extracted.

Semi-solid

In a semi-solid flow cell, the positive and negative electrodes are composed of particles suspended in a carrier liquid. The positive and negative suspensions are stored in separate tanks and pumped through separate pipes into a stack of adjacent reaction chambers, where they are separated by a barrier such as a thin, porous membrane. The approach combines the basic structure of aqueous-flow batteries, which use electrode material suspended in a liquid electrolyte, with the chemistry of lithium-ion batteries in both carbon-free suspensions and slurries with conductive carbon network. The carbon free semi-solid redox flow battery is also sometimes referred to as Solid Dispersion Redox Flow Battery. Dissolving a material changes its chemical behavior significantly. However, suspending bits of solid material preserves the solid's characteristics. The result is a viscous suspension that flows like molasses.

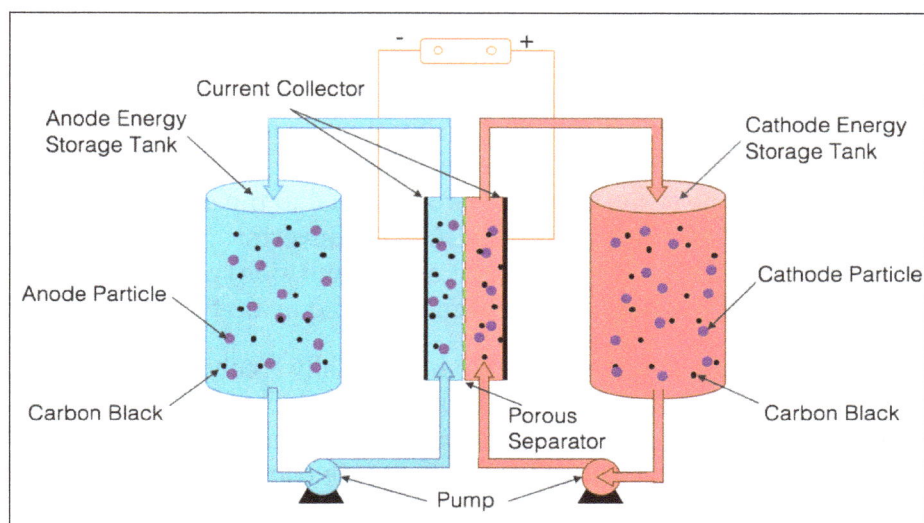

Semi-Solid Flow Battery.

Advantages and Disadvantages

Redox flow batteries, and to a lesser extent hybrid flow batteries, have the advantages of flexible layout (due to separation of the power and energy components), long cycle life (because there are no solid-to-solid phase transitions), quick response times, no need for "equalisation" charging (the overcharging of a battery to ensure all cells have an equal charge) and no harmful emissions. Some types also offer easy state-of-charge determination (through voltage dependence on charge), low maintenance and tolerance to overcharge/overdischarge. They are safe and typically do not contain flammable electrolytes. These technical merits make redox flow batteries a well-suited option for large-scale energy storage.

The two main disadvantages are their low energy density (you need large tanks of electrolyte to store useful amounts of energy) and their low charge and discharge rates (compared to other industrial electrode processes). The latter means that the electrodes and membrane separators need to be large, which increases costs.

Compared to nonreversible fuel cells or electrolyzers using similar electrolytic chemistries, flow batteries generally have somewhat lower efficiency.

Applications

Flow batteries are normally considered for relatively large (1 kWh – 10 MWh) stationary applications. These are for:

- Load balancing: Where the battery is connected to an electrical grid to store excess electrical power during off-peak hours and release electrical power during peak demand periods. The common problem limiting the use of most flow battery chemistries in this application is their low areal power (operating current density) which translates into a high cost of power.

- Storing energy from renewable sources such as wind or solar for discharge during periods of peak demand.

- Peak shaving, where spikes of demand are met by the battery.

- UPS, where the battery is used if the main power fails to provide an uninterrupted supply.

- Power conversion: Because all cells share the same electrolyte/s. Therefore, the electrolyte/s may be charged using a given number of cells and discharged with a different number. Because the voltage of the battery is proportional to the number of cells used the battery can therefore act as a very powerful DC–DC converter. In addition, if the number of cells is continuously changed (on the input and output side) power conversion can also be AC/DC, AC/AC, or DC–AC with the frequency limited by that of the switching gear.

- Electric vehicles: Because flow batteries can be rapidly "recharged" by replacing the electrolyte, they can be used for applications where the vehicle needs to take on energy as fast as a combustion engined vehicle. A common problem found with most RFB chemistries in the EV applications is their low energy density which translated into a short driving range. Flow batteries based on highly soluble halates are a notable exception.

- Stand-alone power system: An example of this is in cellphone base stations where no grid power is available. The battery can be used alongside solar or wind power sources to compensate for their fluctuating power levels and alongside a generator to make the most efficient use of it to save fuel. Currently, flow batteries are being used in solar micro grid applications throughout the Caribbean.

ELECTROCHEMICAL MACHINING

Electrochemical machining (ECM) is a method of removing metal by an electrochemical process. It is normally used for mass production and is used for working extremely hard materials or materials that are difficult to machine using conventional methods. Its use is limited to electrically conductive materials. ECM can cut small or odd-shaped angles, intricate contours or cavities in hard and exotic metals, such as titanium aluminides, Inconel, Waspaloy, and high nickel, cobalt, and rhenium alloys. Both external and internal geometries can be machined.

ECM is often characterized as "reverse electroplating", in that it removes material instead of adding it. It is similar in concept to electrical discharge machining (EDM) in that a high current is passed between an electrode and the part, through an electrolytic material removal process having a negatively charged electrode (anode), a conductive fluid (electrolyte), and a conductive workpiece (cathode); however, in ECM there is no tool wear. The ECM cutting tool is guided along the desired path close to the work but without touching the piece. Unlike EDM, however, no sparks are created. High metal removal rates are possible with ECM, with no thermal or mechanical stresses being transferred to the part, and mirror surface finishes can be achieved.

In the ECM process, a cathode (tool) is advanced into an anode (workpiece). The pressurized electrolyte is injected at a set temperature to the area being cut. The feed rate is the same as the rate of "liquefication" of the material. The gap between the tool and the workpiece varies within 80–800 micrometers (0.003–0.030 in.) As electrons cross the gap, material from the workpiece is dissolved, as the tool forms the desired shape in the workpiece. The electrolytic fluid carries away the metal hydroxide formed in the process.

As far back as 1929, an experimental ECM process was developed by W.Gussef, although it was 1959 before a commercial process was established by the Anocut Engineering Company. B.R. and J.I. Lazarenko are also credited with proposing the use of electrolysis for metal removal.

Much research was done in the 1960s and 1970s, particularly in the gas turbine industry. The rise of EDM in the same period slowed ECM research in the west, although work continued behind the Iron Curtain. The original problems of poor dimensional accuracy and environmentally polluting waste have largely been overcome, although the process remains a niche technique.

The ECM process is most widely used to produce complicated shapes such as turbine blades with good surface finish in difficult to machine materials. It is also widely and effectively used as a deburring process.

In deburring, ECM removes metal projections left from the machining process, and so dulls sharp edges. This process is fast and often more convenient than the conventional methods of deburring by hand or nontraditional machining processes.

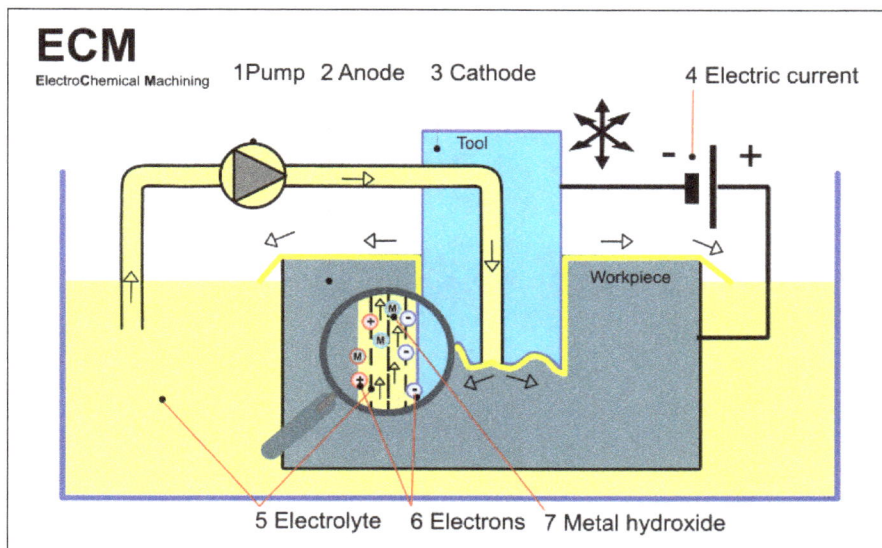

Electrochemical machining principle (ECM) 1 Pump 2 Anode (workpiece)3 Cathode (tool) movable in all direction 4 Electric current 5 Electrolyte 6 Electrons 7 Metal hydroxide.

Advantages

- Complex, concave curvature components can be produced easily by using convex and concave tools.

- Tool wear is zero, same tool can be used for producing infinite number of components.

- No direct contact between tool and work material so there are no forces and residual stresses.

- The surface finish produced is excellent.

- Less heat is generated.

Disadvantages

- The saline (or acidic) electrolyte poses the risk of corrosion to tool, workpiece and equipment.

- Only electrically conductive materials can be machined. High Specific Energy consumption.

- It can not be used for soft material.

Currents Involved

The needed current is proportional to the desired rate of material removal, and the removal rate in mm/minute is proportional to the amps per square mm.

Typical currents range from 0.1 amp per square mm to 5 amps per square mm. Thus, for a small plunge cut of a 1 by 1 mm tool with a slow cut, only 0.1 amps would be needed.

However, for a higher feed rate over a larger area, more current would be used, just like any machining process—removing more material faster takes more power.

Thus, if a current density of 4 amps per square millimeter was desired over a 100×100 mm area, it would take 40,000 amps (and lots of coolant/electrolyte).

Setup and Equipment

An ET 3000 ECM machine by INDEC of Russia.

ECM machines come in both vertical and horizontal types. Depending on the work requirements, these machines are built in many different sizes as well. The vertical machine consists of a base, column, table, and spindle head. The spindle head has a servo-mechanism that automatically advances the tool and controls the gap between the cathode (tool) and the workpiece.

CNC machines of up to six axes are available.

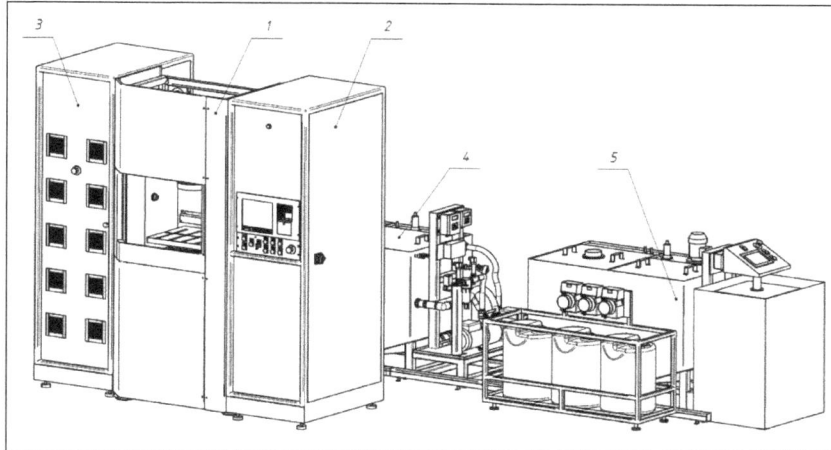

Copper is often used as the electrode material. Brass, graphite, and copper-tungsten are also often used because they are easily machined, they are conductive materials, and they will not corrode.

Applications

Some of the very basic applications of ECM include:

- Die-sinking operations.

- Drilling jet engine turbine blades.

- Multiple hole drilling.

- Machining steam Turbine blades within close limits.

- Micro machining.

- Profiling and conturing.

Similarities between EDM and ECM

- The tool and workpiece are separated by a very small gap, i.e. no contact in between them is made.

- The tool and material must both be conductors of electricity.

- Needs high capital investment.

- Systems consume lots of power.

- A fluid is used as a medium between the tool and the work piece (conductive for ECM and dielectric for EDM).

- The tool is fed continuously towards the workpiece to maintain a constant gap between them (ECM may incorporate intermittent or cyclic, typically partial, tool withdrawal).

Difference between ECM and ECG

- Electrochemical grinding (ECG) is similar to electrochemical machining (ECM) but uses a contoured conductive grinding wheel instead of a tool shaped like the contour of the workpiece.

ELECTROPHORESIS

Electrophoresis is the term used to describe the motion of particles in a gel or fluid within a relatively uniform electric field. Electrophoresis may be used to separate molecules based on charge, size, and binding affinity. The technique is mainly applied to separate and analyze biomolecules, such as DNA, RNA, proteins, nucleic acids, plasmids, and fragments of these macromolecules. Electrophoresis is one of the techniques used to identify source DNA, as in paternity testing and forensic science.

Electrophoresis of anions or negatively charged particles is called anaphoresis. Electrophoresis of cations or positively charged particles is called cataphoresis.

Electrophoresis was first observed in 1807 by Ferdinand Frederic Reuss of Moscow State University, who noticed clay particles migrated in water subjected to a continuous electric field.

How Electrophoresis Works

In electrophoresis, there are two primary factors that control how quickly a particle can move and in what direction. First, the charge on the sample matters. Negatively charged species are attracted to the positive pole of an electric field, while positively charged species are attracted to the negative end. A neutral species may be ionized if the field is strong enough. Otherwise, it doesn't tend to be affected.

The other factor is particle size. Small ions and molecules can move through a gel or liquid much more quickly than larger ones.

While a charged particle is attracted to an opposite charge in an electric field, there are other forces that affect how a molecule moves. Friction and the electrostatic retardation force slow the progress of particles through the fluid or gel. In the case of gel electrophoresis, the concentration of the gel can be controlled to determine the pore size of the gel matrix, which influences mobility. A liquid buffer is also present, which controls the pH of the environment.

As molecules are pulled through a liquid or gel, the medium heats up. This can denature the molecules as well as affect the rate of movement. The voltage is controlled to try to minimize the time required to separate molecules, while maintaining a good separation and keeping the chemical species intact. Sometimes electrophoresis is performed in a refrigerator to help compensate for the heat.

Types of Electrophoresis

Electrophoresis encompasses several related analytical techniques. Examples include:

- Affinity electrophoresis: Affinity electrophoresis is a type of electrophoresis in which particles are separated based on complex formation or biospecific interaction.

- Capillary electrophoresis: Capillary electrophoresis is a type of electrophoresis used to separate ions depending mainly on the atomic radius, charge, and viscosity. As the name suggests, this technique is commonly performed in a glass tube. It yields quick results and a high resolution separation.

- Gel electrophoresis: Gel electrophoresis is a widely used type of electrophoresis in which molecules are separated by movement through a porous gel under the influence of an electrical field. The two main gel materials are agarose and polyacrylamide. Gel electrophoresis is used to separate nucleic acids (DNA and RNA), nucleic acid fragments, and proteins.

- Immunoelectrophoresis: Immunoelectrophoresis is the general name given to a variety of electrophoretic techniques used to characterize and separate proteins based on their reaction to antibodies.

- Electroblotting: Electroblotting is a technique used to recover nucleic acids or proteins following electrophoresis by transferring them onto a membrane. The polymers polyvinylidene fluoride (PVDF) or nitrocellulose are commonly used. Once the specimen has been recovered, it can be further analyzed using stains or probes. A western blot is one form of electroblotting used to detect specific proteins using artificial antibodies.

- Pulsed-field gel electrophoresis: Pulsed-field electrophoresis is used to separate macromolecules, such as DNA, by periodically changing the direction of the electric field applied to a gel matrix. The reason the electric field is changed is because traditional gel electrophoresis is unable to efficiently separate very large molecules that all tend to migrate together. Changing the direction of the electric field gives the molecules additional directions to travel, so they have a path through the gel. The voltage is generally switched between three directions: one running along the axis of the gel and two at 60 degrees to either side. Although the process takes longer than traditional gel electrophoresis, it's better at separating large pieces of DNA.

Isoelectric focusing - Isoelectric focusing (IEF or electrofocusing) is a form of electrophoresis that separates molecules based on different isoelectric points. IEF is most often performed on proteins because their electrical charge depends on pH.

References

- Alotto, P.; Guarnieri, M.; Moro, F. (2014). "Redox Flow Batteries for the storage of renewable energy: a review". Renewable & Sustainable Energy Reviews. 29: 325–335. Doi:10.1016/j.rser.2013.08.001

- Full, fchem.2014.00079, 10.3389: frontiersin.org, Retrieved 30 April, 2019

- Appel, A. M. Et al. "Frontiers, Opportunities, and Challenges in Biochemical and Chemical Catalysis of CO_2 Fixation", Chem. Rev. 2013, vol. 113, 6621-6658. Doi:10.1021/cr300463y

- Electrophoresis-definition-4136322: thoughtco.com, Retrieved 16 January, 2019

- S. Lee et al., chemsuschem 9 (2016) 333-344, Electrode Build-Up of Reducible Metal Composites toward Achievable Electrochemical Conversion of Carbon Dioxidedoi:10.1002/cssc.201501112

- Prabhu, Rahul R. (13 January 2013). "Stationary Fuel Cells Market size to reach 350,000 Shipments by 2022". Renew India Campaign. Retrieved 14 January 2013

- Schmidt-Rohr, K. (2015). "Why Combustions Are Always Exothermic, Yielding About 418 kj per Mole of O_2", J. Chem. Educ. 92: 2094-2099.https://doi.org/10.1021/acs.jchemed.5b00333

Electrolysis

The technique used to drive a non-spontaneous chemical reaction by using a direct electric current is known as electrolysis. Some of the processes that it includes are Castner–Kellner process, Castner process, chloralkali process and Hall-Heroult process. The chapter closely examines these key concepts and processes of electrolysis to provide an extensive understanding of the subject.

Electrolysis is a process by which electrical energy is used to produce a chemical change. Perhaps the most familiar example of electrolysis is the decomposition (breakdown) of water into hydrogen and oxygen by means of an electric current. The same process can be used to decompose compounds other than water. Sodium, chlorine, magnesium, and aluminum are four elements produced commercially by electrolysis.

Principles

The electrolysis of water illustrates the changes that take place when an electric current passes through a chemical compound. Water consists of water molecules, represented by the formula H_2O. In any sample of water, some small fraction of molecules exist in the form of ions, or charged particles. Ions are formed in water when water molecules break apart to form positively charged hydrogen ions and negatively charged hydroxide ions. Chemists describe that process with the following chemical equation:

$$H_2O \rightarrow H^+ + OH^-$$

In order for electrolysis to occur, ions must exist. Seawater can be electrolyzed, for example, because it contains many positively charged sodium ions (Na^+) and negatively charged chloride ions (Cl^-). Any liquid, like seawater, that contains ions is called an electrolyte.

Water is not usually considered an electrolyte because it contains so few hydrogen and hydroxide ions. Normally, only one water molecule out of two billion ionizes. In contrast, sodium chloride (table salt) breaks apart completely when dissolved in water. A salt water solution consists entirely of sodium ions and chloride ions.

In order to electrolyze water, then, one prior step is necessary. Some substance, similar to sodium chloride, must be added to water to make it an electrolyte. The substance that is usually used is sulfuric acid.

The Electrolysis Process

The equipment used for electrolysis of a compound consists of three parts: a source of DC (direct) current; two electrodes; and an electrolyte. A common arrangement consists of a battery (the source of current) whose two poles are attached to two strips of platinum metal (the electrodes), which are immersed in water to which a few drops of sulfuric acid have been added (the electrolyte).

Electrolysis begins when electrical current (a flow of electrons) flows out of one pole of the battery into one electrode, the cathode. Positive hydrogen ions (H^+) in the electrolyte pick up electrons from that electrode and become neutral hydrogen molecules (H_2):

$$2\,H^+ + 2\,e^- \rightarrow H_2$$

(Hydrogen molecules are written as H_2 because they always occur as pairs of hydrogen atoms. The same is true for molecules of oxygen, O_2).

As the electrolysis of water occurs, one can see tiny bubbles escaping from the electrolyte at the cathode. These are bubbles of hydrogen gas.

Commercial Applications

Preparing elements. Electrolysis is used to break down compounds that are very stable. For example, aluminum is a very important metal in modern society. It is used in everything from pots and pans to space shuttles. But the main natural source of aluminum, aluminum oxide, is a very stable compound. A compound that is stable is difficult to break apart. You can't get aluminum out of aluminum oxide just by heating the compound—you need more energy than heat can provide.

Aluminum is prepared by an electrolytic process first discovered in 1886 by a 21-year-old student at Oberlin College in Ohio, Charles Martin Hall. Hall found a way of melting aluminum oxide and then electrolyzing it. Once melted, aluminum oxide forms ions of aluminum and oxygen, which behave in much the same way as hydrogen and hydroxide ions in the previous example. Pure aluminum metal is obtained at the cathode, while oxygen gas bubbles off at the anode. Sodium, chlorine, and magnesium are three other elements obtained commercially by an electrolytic process similar to the Hall process.

Refining of copper. Electrolysis can be used for purposes other than preparing elements. One example is the refining of copper. Very pure copper is often required in the manufacture of electrical equipment. (A purity of 99.999 percent is not unusual.) The easiest way to produce a product of this purity is with electrolysis.

An electrolytic cell for refining copper contains very pure copper at the cathode, impure copper at the anode, and copper sulfate as the electrolyte. When the anode and cathode are connected to a battery, electrons flow into the cathode, where they combine with copper ions (Cu^{2+}) in the electrolyte:

$$Cu^{2+} + 2\,e^- \rightarrow Cu^0$$

Pure copper metal (Cu^0 in the above equation) is formed on the cathode.

At the anode, copper atoms (Cu^0) lose electrons and become copper ions (Cu^{2+}) in the electrolyte:

$$Cu^0 - 2\,e^- \rightarrow Cu^{2+}$$

Overall, the only change that occurs in the cell is that copper atoms from the impure anode become copper ions in the electrolyte. Those copper ions are then plated out on the cathode. Any impurities in the anode are just left behind, and nearly 100 percent pure copper builds up on the cathode.

Electroplating. Another important use of electrolytic cells is in the electroplating of silver, gold, chromium, and nickel. Electroplating produces a very thin coating of these expensive metals on the surfaces of cheaper metals, giving them the appearance and the chemical resistance of the expensive ones.

In silver plating, the object to be plated (a spoon, for example) is used as the cathode. A bar of silver metal is used as the anode. And the electrolyte is a solution of silver cyanide (AgCN). When this arrangement is connected to a battery, electrons flow into the cathode where they combine with silver ions (Ag^+) from the electrolyte to form silver atoms (Ag^0):

$$Ag^+ + 1e^- \rightarrow Ag^0$$

These silver atoms plate out as a thin coating on the cathode—in this case, the spoon. At the anode, silver atoms give up electrons and become silver ions in the electrolyte:

$$Ag^0 - 1e^- \rightarrow Ag^0$$

Silver is cycled, therefore, from the anode to the electrolyte to the cathode, where it is plated out.

ELECTROLYTIC CELL

Voltaic cells use a spontaneous chemical reaction to drive an electric current through an external circuit. These cells are important because they are the basis for the batteries that fuel modern society. But they aren't the only kind of electrochemical cell. It is also possible to construct a cell that does work on a chemical system by driving an electric current through the system. These cells are called electrolytic cells. Electrolysis is used to drive an oxidation-reduction reaction in a direction in which it does not occur spontaneously.

The Electrolysis of Molten NaCl

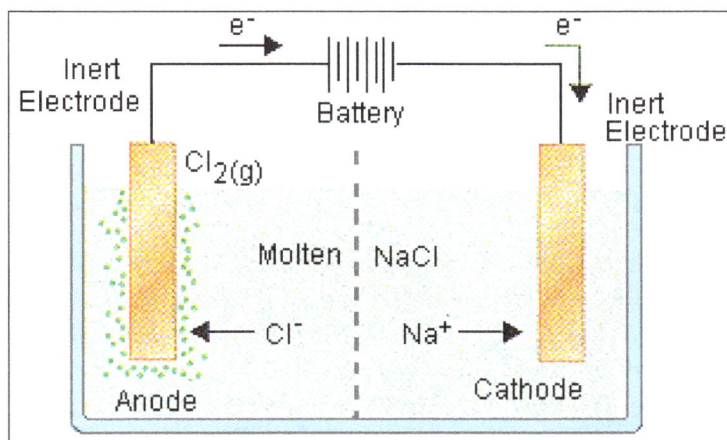

An idealized cell for the electrolysis of sodium chloride is shown in the figure below. A source of direct current is connected to a pair of inert electrodes immersed in molten sodium chloride. Because the salt has been heated until it melts, the Na^+ ions flow toward the negative electrode and the Cl^- ions flow toward the positive electrode.

When Na^+ ions collide with the negative electrode, the battery carries a large enough potential to force these ions to pick up electrons to form sodium metal.

$$\text{Negative electrode (cathode):} \quad Na^+ + e^- \rightarrow Na$$

Cl^- ions that collide with the positive electrode are oxidized to Cl_2 gas, which bubbles off at this electrode.

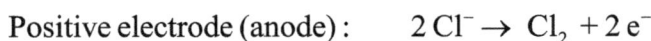

$$\text{Positive electrode (anode):} \quad 2\,Cl^- \rightarrow Cl_2 + 2\,e^-$$

The net effect of passing an electric current through the molten salt in this cell is to decompose sodium chloride into its elements, sodium metal and chlorine gas.

Electrolysis of NaCl :

$$Cathode(-): \quad Na^+ + e^- \rightarrow Na$$
$$Anode(+): \quad 2\,Cl^- \rightarrow Cl_2 + 2\,e^-$$

The potential required to oxidize Cl^- ions to Cl_2 is -1.36 volts and the potential needed to reduce Na^+ ions to sodium metal is -2.71 volts. The battery used to drive this reaction must therefore have a potential of at least 4.07 volts.

This example explains why the process is called electrolysis. Electrolysis literally uses an electric current to split a compound into its elements.

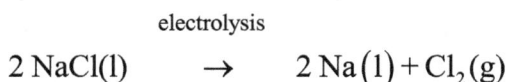

$$2\,NaCl(l) \quad \overset{\text{electrolysis}}{\longrightarrow} \quad 2\,Na(l) + Cl_2(g)$$

This example also illustrates the difference between voltaic cells and electrolytic cells. Voltaic cells use the energy given off in a spontaneous reaction to do electrical work. Electrolytic cells use electrical work as source of energy to drive the reaction in the opposite direction.

The dotted vertical line in the center of the figure represents a diaphragm that keeps the Cl_2 gas produced at the anode from coming into contact with the sodium metal generated at the cathode. The function of this diaphragm can be understood by turning to a more realistic drawing of the commercial Downs cell used to electrolyze sodium chloride shown in the figure below.

Chlorine gas that forms on the graphite anode inserted into the bottom of this cell bubbles through the molten sodium chloride into a funnel at the top of the cell. Sodium metal that forms at the cathode floats up through the molten sodium chloride into a sodium-collecting ring, from which it is periodically drained. The diaphragm that separates the two electrodes is a screen of iron gauze, which prevents the explosive reaction that would occur if the products of the electrolysis reaction came in contact.

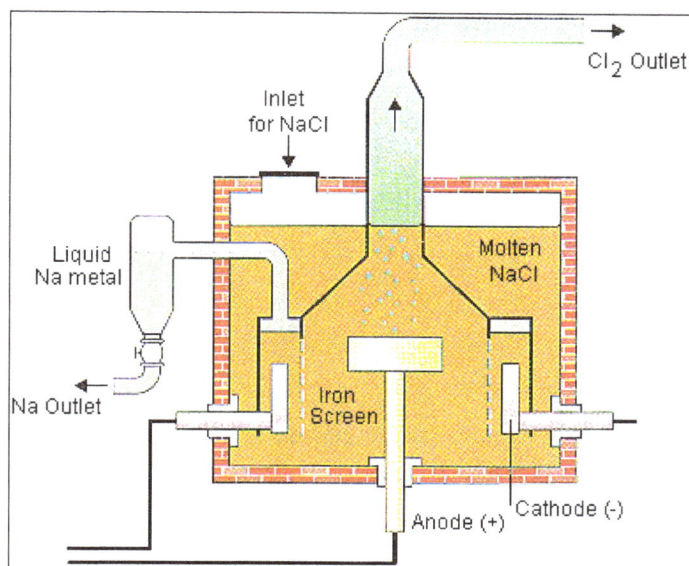

The feed-stock for the Downs cell is a 3:2 mixture by mass of $CaCl_2$ and NaCl. This mixture is used because it has a melting point of 580 °C, whereas pure sodium chloride has to be heated to more than 800 °C before it melts.

The Electrolysis of Aqueous NaCl

The figure below shows an idealized drawing of a cell in which an aqueous solution of sodium chloride is electrolyzed.

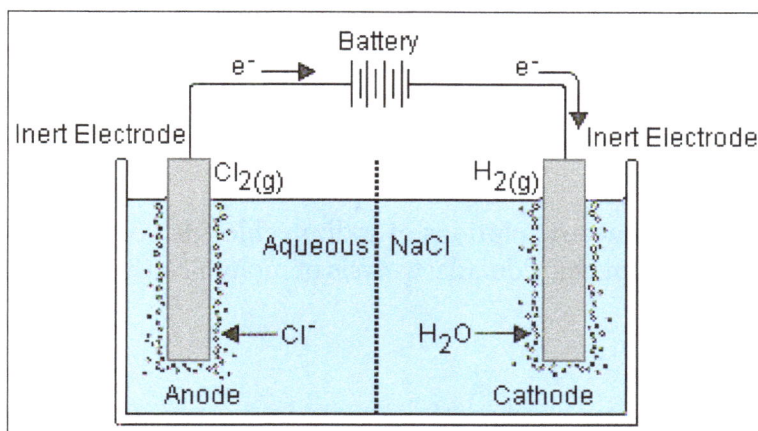

Once again, the Na^+ ions migrate toward the negative electrode and the Cl^- ions migrate toward the positive electrode. But, now there are two substances that can be reduced at the cathode: Na^+ ions and water molecules.

Cathode(-):

$$Na^+ + e^- \rightarrow Na \qquad\qquad E°_{red} = -2.71\,V$$

$$2\,H_2O + 2\,e^- \rightarrow H_2 + 2\,OH^- \qquad E°_{red} = -0.83\,V$$

Because it is much easier to reduce water than Na^+ ions, the only product formed at the cathode is hydrogen gas.

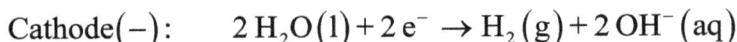

Cathode$(-)$: $2\,H_2O(l) + 2\,e^- \rightarrow H_2(g) + 2\,OH^-(aq)$

There are also two substances that can be oxidized at the anode: Cl^- ions and water molecules.

Anode$(+)$:

$$2\,Cl^- \rightarrow Cl_2 + 2\,e^- \qquad E^\circ_{ox} = -1.36\ V$$

$$2\,H_2O \rightarrow O_2 + 4\,H^+ + 4\,e^- \qquad E^\circ_{ox} = -1.23\ V$$

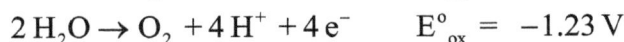

The standard-state potentials for these half-reactions are so close to each other that we might expect to see a mixture of Cl_2 and O_2 gas collect at the anode. In practice, the only product of this reaction is Cl_2.

Anode$(+)$: $2\,Cl^- \rightarrow Cl_2 + 2\,e^-$

At first glance, it would seem easier to oxidize water (E°_{ox} = -1.23 volts) than Cl^- ions (E°_{ox} = -1.36 volts). It is worth noting, however, that the cell is never allowed to reach standard-state conditions. The solution is typically 25% NaCl by mass, which significantly decreases the potential required to oxidize the Cl^- ion. The pH of the cell is also kept very high, which decreases the oxidation potential for water. The deciding factor is a phenomenon known as overvoltage, which is the extra voltage that must be applied to a reaction to get it to occur at the rate at which it would occur in an ideal system.

Under ideal conditions, a potential of 1.23 volts is large enough to oxidize water to O_2 gas. Under real conditions, however, it can take a much larger voltage to initiate this reaction. (The overvoltage for the oxidation of water can be as large as 1 volt.) By carefully choosing the electrode to maximize the overvoltage for the oxidation of water and then carefully controlling the potential at which the cell operates, we can ensure that only chlorine is produced in this reaction.

In summary, electrolysis of aqueous solutions of sodium chloride doesn't give the same products as electrolysis of molten sodium chloride. Electrolysis of molten NaCl decomposes this compound into its elements.

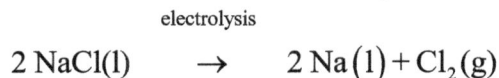

$$2\,NaCl(l) \xrightarrow{\text{electrolysis}} 2\,Na(l) + Cl_2(g)$$

Electrolysis of aqueous NaCl solutions gives a mixture of hydrogen and chlorine gas and an aqueous sodium hydroxide solution.

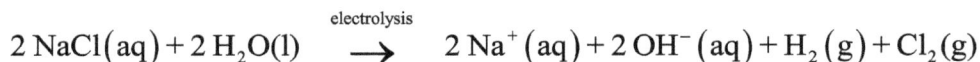

$$2\,NaCl(aq) + 2\,H_2O(l) \xrightarrow{\text{electrolysis}} 2\,Na^+(aq) + 2\,OH^-(aq) + H_2(g) + Cl_2(g)$$

Because the demand for chlorine is much larger than the demand for sodium, electrolysis of aqueous sodium chloride is a more important process commercially. Electrolysis of an aqueous NaCl solution has two other advantages. It produces H_2 gas at the cathode, which can be collected and sold. It also produces NaOH, which can be drained from the bottom of the electrolytic cell and sold.

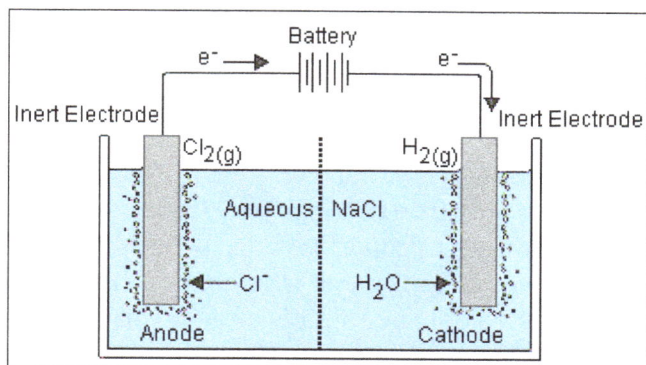

The dotted vertical line in the above figure represents a diaphragm that prevents the Cl_2 produced at the anode in this cell from coming into contact with the NaOH that accumulates at the cathode. When this diaphragm is removed from the cell, the products of the electrolysis of aqueous sodium chloride react to form sodium hypo-chlorite, which is the first step in the preparation of hypochlorite bleaches, such as Chlorox.

$$Cl_2(g) + 2OH^-(aq) \rightarrow Cl^-(aq) + OCl^-(aq) + H_2O(l)$$

HOFMANN VOLTAMETER

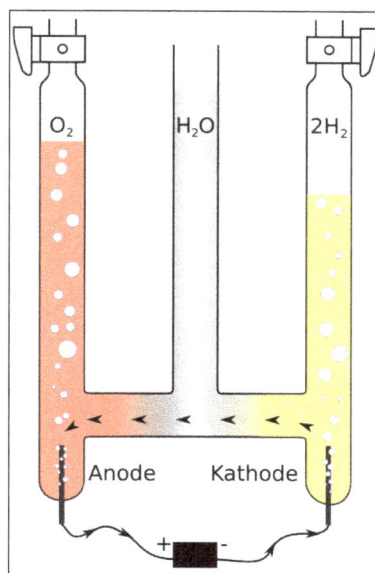

Hofmann voltameter.

A Hofmann voltameter is an apparatus for electrolysing water, invented by August Wilhelm von Hofmann in 1866. It consists of three joined upright cylinders, usually glass. The inner cylinder is open at the top to allow addition of water and an ionic compound to improve conductivity, such as a small amount of sulfuric acid. A platinum electrode is placed inside the bottom of each of the two side cylinders, connected to the positive and negative terminals of a source of electricity. When current is run through Hofmann's voltameter, gaseous oxygen forms at the anode and gaseous hydrogen at the cathode. Each gas displaces water and collects at the top of the two outer tubes.

The name 'voltameter' was coined by Daniell, who shortened Faraday's original name of "volta-electrometer". Hofmann voltameters are no longer used as electrical measuring devices. However, before the invention of the ammeter, voltameters were often used to measure direct current, since current through a voltameter with iron or copper electrodes electroplates the cathode with an amount of metal from the anode directly proportional to the total coulombs of charge transferred (Faraday's law of electrolysis). The modern name is "electrochemical coulometer". Although the correct spelling of Hofmann contains only one 'f', it is often incorrectly depicted as Hoffmann.

Uses

The amount of electricity that has passed through the system can then be determined by weighing the cathode. Thomas Edison used voltameters as electricity meters. (A Hofmann voltameter cannot be used to weigh electric current in this fashion, as the platinum electrodes are too inert for plating).

A Hofmann voltameter is often used as a demonstration of stoichiometric principles, as the two-to-one ratio of the volumes of hydrogen and oxygen gas produced by the apparatus illustrates the chemical formula of water, H_2O. However, this is only true if oxygen and hydrogen gases are assumed to be diatomic. If hydrogen gas were monatomic and oxygen diatomic, the gas volume ratio would be 4:1. The volumetric composition of water is the ratio by volume of hydrogen to oxygen present. This value is 2:1 experimentally; this value is determined using Hofmann's water voltameter.

FARADAYS LAWS OF ELECTROLYSIS

Faraday's laws of electrolysis are quantitative (mathematical) relationships that describe the above two phenomena.

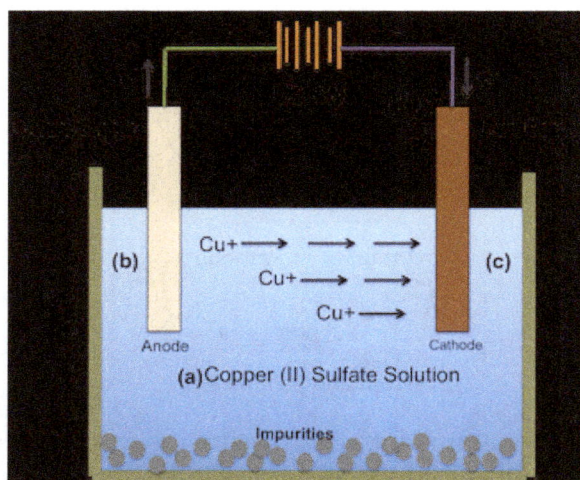

Faraday's First and Second Laws of Electrolysis.

Faraday's First Law of Electrolysis

It is clear that the flow of current through the external battery circuit fully depends upon how many electrons get transferred from negative electrode or cathode to positive metallic ion or cations. If the cations have valency of two like Cu^{++} then for every cation, there would be two electrons transferred from cathode to cation. We know that every electron has negative electrical charge – 1.602 × 10⁻¹⁹ Coulombs and say it is – e. So for disposition of every Cu atom on the cathode, there would be – 2.e charge transfers from cathode to cation.

Now say for t time there would be total n number of copper atoms deposited on the cathode, so total charge transferred, would be – 2.n.e Coulombs. Mass m of the deposited copper is obviously a function of the number of atoms deposited. So, it can be concluded that the mass of the deposited copper is directly proportional to the quantity of electrical charge that passes through the electrolyte. Hence mass of deposited copper m ∝ Q quantity of electrical charge passes through the electrolyte.

Faraday's First Law of Electrolysis states that the chemical deposition due to the flow of current through an electrolyte is directly proportional to the quantity of electricity (coulombs) passed through it. i.e. mass of chemical deposition:

$$m \propto \text{Quantity of electricity}, Q \Rightarrow m = Z, Q$$

Where, Z is a constant of proportionality and is known as electro-chemical equivalent of the substance.

If we put Q = 1 coulombs in the above equation, we will get Z = m which implies that electrochemical equivalent of any substance is the amount of the substance deposited on the passing of 1 coulomb through its solution. This constant of the passing of electrochemical equivalent is generally expressed in terms of milligrams per coulomb or kilogram per coulomb.

Faraday's Second Law of Electrolysis

So far we have learned that the mass of the chemical, deposited due to electrolysis is proportional to the quantity of electricity that passes through the electrolyte. The mass of the chemical, deposited due to electrolysis is not only proportional to the quantity of electricity passes through the electrolyte, but it also depends upon some other factor. Every substance will have its own atomic weight. So for the same number of atoms, different substances will have different masses.

Again, how many atoms deposited on the electrodes also depends upon their number of valency. If valency is more, then for the same amount of electricity, the number of deposited atoms will be less whereas if valency is less, then for the same quantity of electricity, morenumber of atoms to be deposited.

So, for the same quantity of electricity or charge passes through different electrolytes, the mass of deposited chemical is directly proportional to its atomic weight and inversely proportional to its valency.

Faraday's second law of electrolysis states that, when the same quantity of electricity is passed through several electrolytes, the mass of the substances deposited are proportional to their respective chemical equivalent or equivalent weight.

Chemical Equivalent or Equivalent Weight

The chemical equivalent or equivalent weight of a substance can be determined by Faraday's laws of electrolysis, and it is defined as the weight of that subtenancy which will combine with or displace the unit weight of hydrogen.

The chemical equivalent of hydrogen is, thus, unity. Since valency of a substance is equal to the number of hydrogen atoms, which it can replace or with which it can combine, the chemical equivalent of a substance, therefore may be defined as the ratio of its atomic weight to its valency.

$$\text{Thus chemical equivalent} = \frac{\text{Atomic weight}}{V\text{alency}}$$

Who Invented Faraday's Laws of Electrolysis?

Faraday's Laws of Electrolysis were published by Michael Faraday in 1834. Michael Faraday was also responsible.

Michael Faraday.

As well as discovering these laws of electrolysis, Michael Faraday is also responsible for popularizing terminologies such as electrodes, ions, anodes, and cathodes.

PRINCIPLE OF ELECTROLYSIS OF COPPER SULFATE ELECTROLYTE

An electrolyte is such a chemical whose atoms are tightly bonded together, by ionic bonds but when we dissolve it in water, its molecules split up into positive, and negative ions. The positively charged ions are referred as cations whereas negatively charged ions are referred as anions. Both cations and anions move freely in the solution.

Principle of Electrolysis

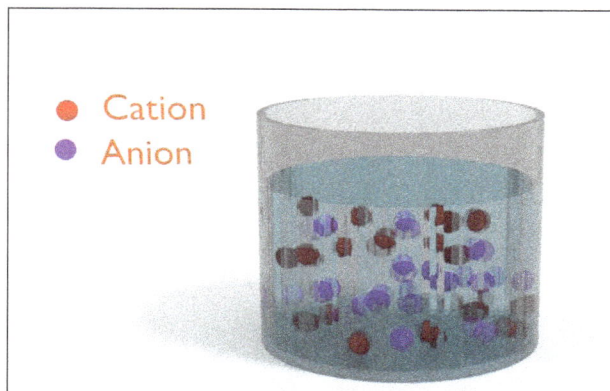

Now we will immerse two metal rods in the solution and we will apply an electrical potential difference between the rods externally by a battery.

In ionic bonds, one atom loses its valence electrons and another atom gains electrons. As a result, one atom becomes positively charged ion and another atom becomes a negative ion. Due to opposite charge both attract each other and form a bonding between them called the ionic bond. In ionic bond, the force acting between the ions is Coulombic force which is inversely proportional to the permittivity of the medium. The relative permittivity of water is 80 at 20 °C. So, when any ionic bonded chemical is dissolved in water, the bonding strength between ions becomes much weaker and hence its molecules split into cations and anions moving freely in the solution.

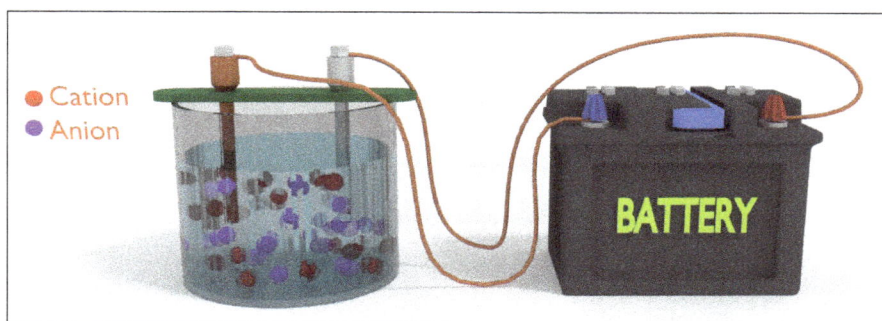

These partly immersed rods are technically referred as electrodes. The electrode connected with negative terminal of the battery is known as cathode and the electrode connected with positive terminal of the battery is known as anode.The freely moving positively charged cations are attracted by cathode and negatively charged anions are attracted by anode. In cathode, the positive cations take electrons from negative cathode and in anode, negative anions give electrons to the positive anode. For continually taking and giving electrons in cathode and anode respectively, there must be flow of electrons in the external circuit of the electrolytic. That means, current continues to circulate around the closed loop created by battery, electrolytic and electrodes. This is the most basic principle of electrolysis.

Electrolysis of Copper Sulfate

Whenever copper sulfate or $CuSO_4$ is added to water, it gets dissolved in the water. As $CuSO_4$ is an electrolyte, it splits into Cu^{++} (cation) and SO_4^{--} (anion) ions and move freely in the solution.

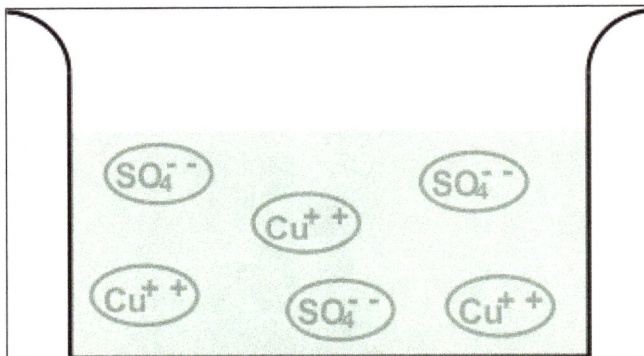

Now we will immerse two copper electrodes in that solution.

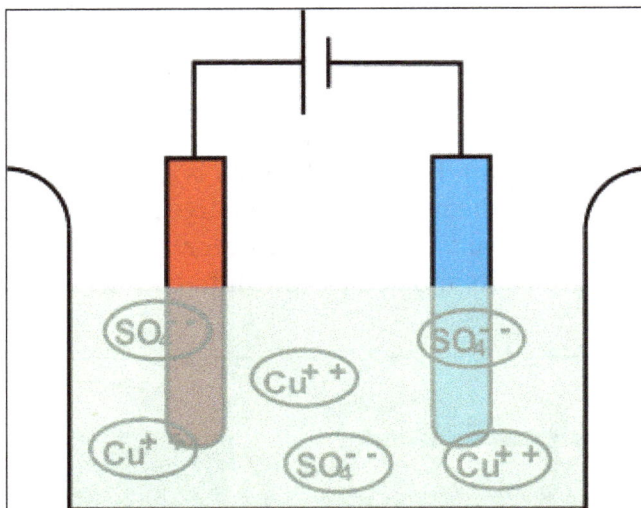

The Cu^{++} ions (cation) will be attracted towards cathode i.e. the electrode connected to the negative terminal of the battery. On reaching on the cathode, each Cu^{++} ion will take electrons from it and becomes neutral copper atoms.

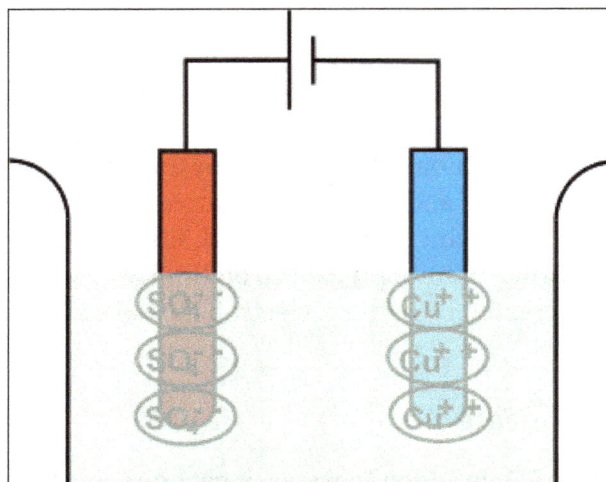

Similarly the SO_4^{--} (anion) ions will be attracted by anode i.e. the electrode connected to the positive terminal of the battery. So SO_4^{--} ions will move towards anode where they give up two electrons and become SO4 radical.

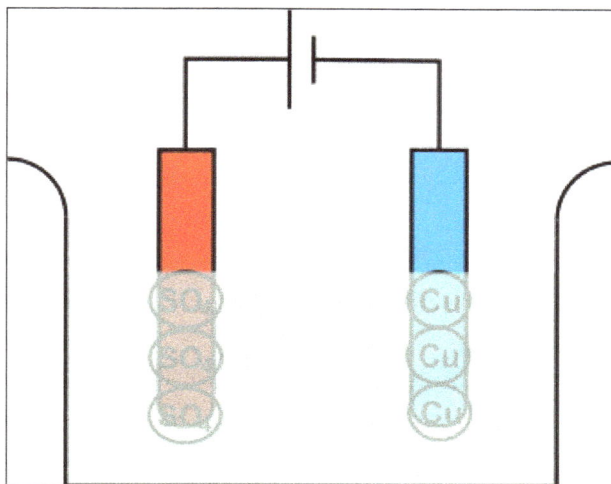

But since SO_4 radical can not exist in the electrical neutral state, it will attack copper anode and will form copper sulfate.

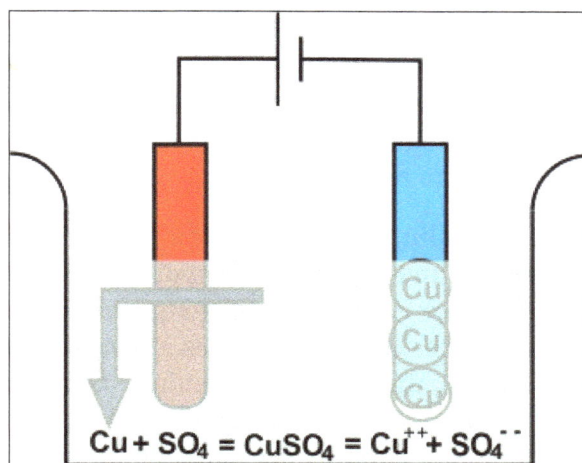

$$Cu + SO_4 = CuSO_4 = Cu^{++} + SO_4^{--}$$

In the above process, after taking electrons the neutral copper atoms get deposited on the cathode. At the same time, SO_4 reacts with copper anode and becomes $CuSO_4$ but in water it can not exist as single molecules instead of that $CuSO_4$ will split into Cu^{++}, SO_4^{--} and dissolve in water. So it can be concluded that, during electrolysis of copper sulfate with copper electrodes, copper is deposited on cathode and same amount of copper is removed from anode. If during electrolysis of copper sulfate, we use carbon electrode instead of copper or other metal electrodes, then electrolysis reactions will be little bit different. Actually SO_4 can not react with carbon and in this case the SO_4 will react with water of the solution and will form sulfuric acid and liberate oxygen.

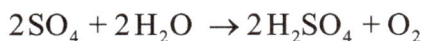

$$2SO_4 + 2H_2O \rightarrow 2H_2SO_4 + O_2$$

The process described above is known as electrolysis.

ANODIZING

These carabiners have an anodized aluminium surface that has been dyed; they are made in many colors.

Anodising is an electrolytic passivation process used to increase the thickness of the natural oxide layer on the surface of metal parts.

The process is called anodising because the part to be treated forms the anode electrode of an electrolytic cell. Anodising increases resistance to corrosion and wear, and provides better adhesion for paint primers and glues than bare metal does. Anodic films can also be used for a number of cosmetic effects, either with thick porous coatings that can absorb dyes or with thin transparent coatings that add interference effects to reflected light.

Anodising is also used to prevent galling of threaded components and to make dielectric films for electrolytic capacitors. Anodic films are most commonly applied to protect aluminium alloys, although processes also exist for titanium, zinc, magnesium, niobium, zirconium, hafnium, and tantalum. Iron or carbon steel metal exfoliates when oxidized under neutral or alkaline microelectrolytic conditions; i.e., the iron oxide (actually ferric hydroxide or hydrated iron oxide, also known as rust) forms by anoxic anodic pits and large cathodic surface, these pits concentrate anions such as sulfate and chloride accelerating the underlying metal to corrosion. Carbon flakes or nodules in iron or steel with high carbon content (high-carbon steel, cast iron) may cause an electrolytic potential and interfere with coating or plating. Ferrous metals are commonly anodized electrolytically in nitric acid or by treatment with red fuming nitric acid to form hard black Iron(II,III) oxide. This oxide remains conformal even when plated on wire and the wire is bent.

Anodizing changes the microscopic texture of the surface and the crystal structure of the metal near the surface. Thick coatings are normally porous, so a sealing process is often needed to achieve corrosion resistance. Anodized aluminium surfaces, for example, are harder than aluminium but have low to moderate wear resistance that can be improved with increasing thickness or by applying suitable sealing substances. Anodic films are generally much stronger and more adherent than most types of paint and metal plating, but also more brittle. This makes them less likely to crack and peel from aging and wear, but more susceptible to cracking from thermal stress.

Aluminum

Coloured anodized aluminum key blanks.

Aluminium alloys are anodized to increase corrosion resistance and to allow dyeing (coloring), improved lubrication, or improved adhesion. However, anodizing does not increase the strength of the aluminium object. The anodic layer is non-conductive.

When exposed to air at room temperature, or any other gas containing oxygen, pure aluminum self-passivates by forming a surface layer of amorphous aluminum oxide 2 to 3 nm thick, which provides very effective protection against corrosion. Aluminum alloys typically form a thicker oxide layer, 5–15 nm thick, but tend to be more susceptible to corrosion. Aluminium alloy parts are anodized to greatly increase the thickness of this layer for corrosion resistance. The corrosion resistance of aluminium alloys is significantly decreased by certain alloying elements or impurities: copper, iron, and silicon, so 2000-, 4000-, and 6000-series Al alloys tend to be most susceptible.

Although anodizing produces a very regular and uniform coating, microscopic fissures in the coating can lead to corrosion. Further, the coating is susceptible to chemical dissolution in the presence of high- and low-pH chemistry, which results in stripping the coating and corrosion of the substrate. To combat this, various techniques have been developed either to reduce the number of fissures, to insert more chemically stable compounds into the oxide, or both. For instance, sulfuric-anodized articles are normally sealed, either through hydro-thermal sealing or precipitating sealing, to reduce porosity and interstitial pathways that allow corrosive ion exchange between the surface and the substrate. Precipitating seals enhance chemical stability but are less effective in eliminating ion exchange pathways. Most recently, new techniques to partially convert the amorphous oxide coating into more stable micro-crystalline compounds have been developed that have shown significant improvement based on shorter bond lengths.

Some aluminium aircraft parts, architectural materials, and consumer products are anodized. Anodized aluminium can be found on MP3 players, smartphones, multi-tools, flashlights, cookware, cameras, sporting goods, firearms, window frames, roofs, in electrolytic capacitors, and on many other products both for corrosion resistance and the ability to retain dye. Although anodizing only has moderate wear resistance, the deeper pores can better retain a lubricating film than a smooth surface would.

Anodized coatings have a much lower thermal conductivity and coefficient of linear expansion than aluminium. As a result, the coating will crack from thermal stress if exposed to temperatures above 80 °C (353 K). The coating can crack, but it will not peel. The melting point of aluminium oxide is 2050 °C (2323 K), much higher than pure aluminium's 658 °C (931 K). This and the non-conductivity of aluminum oxide can make welding more difficult.

In typical commercial aluminium anodizing processes, the aluminium oxide is grown down into the surface and out from the surface by equal amounts. So anodizing will increase the part dimensions on each surface by half the oxide thickness. For example, a coating that is 2 µm thick will increase the part dimensions by 1 µm per surface. If the part is anodized on all sides, then all linear dimensions will increase by the oxide thickness. Anodized aluminium surfaces are harder than aluminium but have low to moderate wear resistance, although this can be improved with thickness and sealing.

Process

The anodized aluminium layer is grown by passing a direct current through an electrolytic solution, with the aluminium object serving as the anode (the positive electrode). The current releases hydrogen at the cathode (the negative electrode) and oxygen at the surface of the aluminium anode, creating a build-up of aluminium oxide. Alternating current and pulsed current is also possible but rarely used. The voltage required by various solutions may range from 1 to 300 V DC, although most fall in the range of 15 to 21 V. Higher voltages are typically required for thicker coatings formed in sulfuric and organic acid. The anodizing current varies with the area of aluminium being anodized and typically ranges from 30 to 300 A/m² (2.8 to 28 A/ft²).

Aluminium anodizing is usually performed in an acid solution, which slowly dissolves the aluminium oxide. The acid action is balanced with the oxidation rate to form a coating with nanopores, 10–150 nm in diameter. These pores are what allow the electrolyte solution and current to reach the aluminium substrate and continue growing the coating to greater thickness beyond what is produced by autopassivation. However, these same pores will later permit air or water to reach the substrate and initiate corrosion if not sealed. They are often filled with colored dyes and corrosion inhibitors before sealing. Because the dye is only superficial, the underlying oxide may continue to provide corrosion protection even if minor wear and scratches may break through the dyed layer.

Conditions such as electrolyte concentration, acidity, solution temperature, and current must be controlled to allow the formation of a consistent oxide layer. Harder, thicker films tend to be produced by more dilute solutions at lower temperatures with higher voltages and currents. The film thickness can range from under 0.5 micrometers for bright decorative work up to 150 micrometers for architectural applications.

Dual-finishing

Anodizing can be performed in combination with chromate conversion coating. Each process provides corrosion resistance, with anodizing offering a significant advantage when it comes to ruggedness or physical wear resistance. The reason for combining the processes can vary, however, the significant difference between anodizing and chromate conversion coating is the electrical conductivity of the films produced. Although both stable compounds, chromate conversion coating has a greatly increased electrical conductivity. Applications where this may be useful are varied, however, the issue of grounding components as part of a larger system is an obvious one.

The dual finishing process uses the best each process has to offer, anodizing with its hard wear resistance and chromate conversion coating with its electrical conductivity.

The process steps can typically involve chromate conversion coating the entire component, followed by a masking of the surface in areas where the chromate coating must remain intact. Beyond that, the chromate coating is then dissolved in unmasked areas. The component can then be anodized, with anodizing taking to the unmasked areas. The exact process will vary dependent on service provider, component geometry and required outcome. It helps to protect aluminium article.

Other Widely used Specifications

The most widely used anodizing specification in the US is a U.S. military spec, MIL-A-8625, which defines three types of aluminium anodizing. Type I is chromic acid anodising, Type II is sulfuric acid anodizing, and Type III is sulfuric acid hard anodizing. Other anodizing specifications include more MIL-SPECs (e.g., MIL-A-63576), aerospace industry specs by organizations such as SAE, ASTM, and ISO (e.g., AMS 2469, AMS 2470, AMS 2471, AMS 2472, AMS 2482, ASTM B580, ASTM D3933, ISO 10074, and BS 5599), and corporation-specific specs (such as those of Boeing, Lockheed Martin, Airbus and other large contractors). AMS 2468 is obsolete. None of these specifications define a detailed process or chemistry, but rather a set of tests and quality assurance measures which the anodized product must meet. BS 1615 provides guidance in the selection of alloys for anodizing. For British defense work, a detailed chromic and sulfuric anodizing processes are described by DEF STAN 03-24/3 and DEF STAN 03-25/3 respectively.

Chromic Acid (Type I)

The oldest anodizing process uses chromic acid. It is widely known as the Bengough-Stuart process but, due to the safety regulations regarding air quality control, is not preferred by vendors when the additive material associated with type II doesn't break tolerances. In North America it is known as Type I because it is so designated by the MIL-A-8625 standard, but it is also covered by AMS 2470 and MIL-A-8625 Type IB. In the UK it is normally specified as Def Stan 03/24 and used in areas that are prone to come into contact with propellants etc. There are also Boeing and Airbus standards. Chromic acid produces thinner, 0.5 μm to 18 μm (0.00002" to 0.0007") more opaque films that are softer, ductile, and to a degree self-healing. They are harder to dye and may be applied as a pretreatment before painting. The method of film formation is different from using sulfuric acid in that the voltage is ramped up through the process cycle.

Sulfuric Acid (Type II and III)

Sulfuric acid is the most widely used solution to produce anodized coating. Coatings of moderate thickness 1.8 μm to 25 μm (0.00007" to 0.001") are known as Type II in North America, as named by MIL-A-8625, while coatings thicker than 25 μm (0.001") are known as Type III, hardcoat, hard anodizing, or engineered anodizing. Very thin coatings similar to those produced by chromic anodizing are known as Type IIB. Thick coatings require more process control, and are produced in a refrigerated tank near the freezing point of water with higher voltages than the thinner coatings. Hard anodizing can be made between 13 and 150 μm (0.0005" to 0.006") thick. Anodizing thickness increases wear resistance, corrosion resistance, ability to retain lubricants and PTFE coat-ings, and electrical and thermal insulation. Standards for thin (Soft/Standard) sulfuric anodizing are given by MIL-A-8625 Types II and IIB, AMS 2471 (undyed), and AMS 2472 (dyed), BS EN ISO 12373/1 (decorative), BS 3987 (Architectural). Standards for thick sulfuric anodizing are given by MIL-A-8625 Type III, AMS 2469, BS ISO 10074, BS EN 2536 and the obsolete AMS 2468 and DEF STAN 03-26/1.

Organic Acid

Anodizing can produce yellowish integral colors without dyes if it is carried out in weak acids with high voltages, high current densities, and strong refrigeration. Shades of color are restricted to a range which includes pale yellow, gold, deep bronze, brown, grey, and black. Some advanced variations can produce a white coating with 80% reflectivity. The shade of color produced is sensitive to variations in the metallurgy of the underlying alloy and cannot be reproduced consistently.

Anodizing in some organic acids, for example malic acid, can enter a 'runaway' situation, in which the current drives the acid to attack the aluminum far more aggressively than normal, resulting in huge pits and scarring. Also, if the current or voltage are driven too high, 'burning' can set in; in this case the supplies act as if nearly shorted and large, uneven and amorphous black regions develop.

Integral colour anodizing is generally done with organic acids, but the same effect has been produced in laboratory with very dilute sulfuric acid. Integral color anodizing was originally performed with oxalic acid, but sulfonated aromatic compounds containing oxygen, particularly sulfosalicylic acid, have been more common since the 1960s. Thicknesses up to 50 μm can be achieved. Organic acid anodizing is called Type IC by MIL-A-8625.

Phosphoric Acid

Anodizing can be carried out in phosphoric acid, usually as a surface preparation for adhesives. This is described in standard ASTM D3933.

Borate and Tartrate Baths

Anodizing can also be performed in borate or tartrate baths in which aluminium oxide is insoluble. In these processes, the coating growth stops when the part is fully covered, and the thickness is linearly related to the voltage applied. These coatings are free of pores, relative to the sulfuric and chromic acid processes. This type of coating is widely used to make electrolytic capacitors, because the thin aluminium films (typically less than 0.5 μm) would risk being pierced by acidic processes.

Plasma Electrolytic Oxidation

Plasma electrolytic oxidation is a similar process, but where higher voltages are applied. This causes sparks to occur, and results in more crystalline/ceramic type coatings.

Other Metals

Magnesium

Magnesium is anodized primarily as a primer for paint. A thin (5 μm) film is sufficient for this. Thicker coatings of 25 μm and up can provide mild corrosion resistance when sealed with oil, wax, or sodium silicate. Standards for magnesium anodizing are given in AMS 2466, AMS 2478, AMS 2479, and ASTM B893.

Niobium

Niobium anodizes in a similar fashion to titanium with a range of attractive colors being formed by interference at different film thicknesses. Again the film thickness is dependent on the anodizing voltage. Uses include jewelry and commemorative coins.

Tantalum

Tantalum anodizes in a similar fashion to titanium and niobium with a range of attractive colors being formed by interference at different film thicknesses. Again the film thickness is dependent on the anodizing voltage and typically ranges from 18 to 23 Angstroms per volt depending on electrolyte and temperature. Uses include tantalum capacitors.

Titanium

Selected colors achievable through anodization of titanium.

An anodized oxide layer has a thickness in the range of 30 nanometers (1.2×10^{-6} in) to several micrometers. Standards for titanium anodizing are given by AMS 2487 and AMS 2488.

AMS 2488 Type III anodizing of titanium generates an array of different colors without dyes, for which it is sometimes used in art, costume jewelry, body piercing jewelry and wedding rings. The color formed is dependent on the thickness of the oxide (which is determined by the anodizing voltage); it is caused by the interference of light reflecting off the oxide surface with light traveling through it and reflecting off the underlying metal surface. AMS 2488 Type II anodizing produces a thicker matte gray finish with higher wear resistance.

Zinc

Zinc is rarely anodized, but a process was developed by the International Lead Zinc Research Organization and covered by MIL-A-81801. A solution of ammonium phosphate, chromate and fluoride with voltages of up to 200 V can produce olive green coatings up to 80 μm thick. The coatings are hard and corrosion resistant.

Zinc or galvanized steel can be anodized at lower voltages (20–30 V) as well as using direct currents from silicate baths containing varying concentration of sodium silicate, sodium hydroxide, borax, sodium nitrite and nickel sulphate.

Dyeing

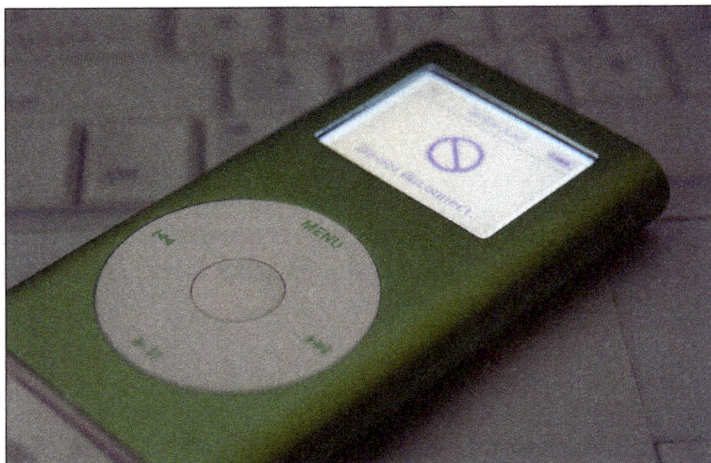

Colored iPod Mini cases are dyed following anodizing and before thermal sealing.

The most common anodizing processes, for example sulfuric acid on aluminium, produce a porous surface which can accept dyes easily. The number of dye colors is almost endless; however, the colors produced tend to vary according to the base alloy. The most common colors in industry, due to them being relatively cheap, are yellow, green, blue, black, orange, purple and red. Though some may prefer lighter colors, in practice they may be difficult to produce on certain alloys such as high-silicon casting grades and 2000-series aluminium-copper alloys. Another concern is the "lightfastness" of organic dyestuffs—some colors (reds and blues) are particularly prone to fading. Black dyes and gold produced by inorganic means (ferric ammonium oxalate) are more lightfast. Dyed anodizing is usually sealed to reduce or eliminate dye bleed out.

Alternatively, metal (usually tin) can be electrolytically deposited in the pores of the anodic coating to provide colors that are more lightfast. Metal dye colors range from pale champagne to black. Bronze shades are commonly used for architectural metals.

Alternatively the color may be produced integral to the film. This is done during the anodizing process using organic acids mixed with the sulfuric electrolyte and a pulsed current.

Splash effects are created by dying the unsealed porous surface in lighter colors and then splashing darker color dyes onto the surface. Aqueous and solvent-based dye mixtures may also be alternately applied since the colored dyes will resist each other and leave spotted effects.

Sealing

Acidic anodizing solutions produce pores in the anodized coating. These pores can absorb dyes and retain lubricants, but are also an avenue for corrosion. When lubrication properties are not critical, they are usually sealed after dyeing to increase corrosion resistance and dye retention. Long immersion in boiling-hot deionized water or steam is the simplest sealing process, although it is not completely effective and reduces abrasion resistance by 20%. The oxide is converted into its hydrated form, and the resulting swelling reduces the porosity of the surface. Cold sealing, where the pores are closed by impregnation of a sealant in a room-temperature bath, is more popular due to energy savings. Coatings sealed in this method are not suitable for adhesive bonding. Teflon, nickel acetate, cobalt acetate, and hot sodium or potassium dichromate seals are commonly used. MIL-A-8625 requires sealing for thin coatings (Types I and II) and allows it as an option for thick ones (Type III).

Cleaning

Anodized aluminium surfaces that are not regularly cleaned are susceptible to panel edge staining, a unique type of surface staining that can affect the structural integrity of the metal.

Environmental Impact

Anodizing is one of the more environmentally friendly metal finishing processes. With the exception of organic (aka integral color) anodizing, the by-products contain only small amounts of heavy metals, halogens, or volatile organic compounds. Integral color anodizing produces no VOCs, heavy metals, or halogens as all of the byproducts found in the effluent streams of other processes come from their dyes or plating materials. The most common anodizing effluents, aluminium hydroxide and aluminium sulfate, are recycled for the manufacturing of alum, baking powder, cosmetics, newsprint and fertilizer or used by industrial wastewater treatment systems.

Mechanical Considerations

Anodizing will raise the surface, since the oxide created occupies more space than the base metal converted. This will generally not be of consequence except where there are tight tolerances. If so, the thickness of the anodizing layer has to be taken into account when choosing the machining dimension. A general practice on engineering drawing is to specify that "dimensions apply after all surface finishes". This will force the machine shop to take into account the anodization thickness when performing final machining of the mechanical part prior to anodization. Also in the case of small holes threaded to accept screws, anodizing may cause the screws to bind, thus the threaded holes may need to be chased with a tap to restore the original dimensions. Alternatively, special oversize taps may be used to precompensate for this growth. In the case of unthreaded holes that accept fixed-diameter pins or rods, a slightly oversized hole to allow for the dimension change may be appropriate. Depending on the alloy and thickness of the anodized coating, the same may have a significantly negative effect on fatigue life. On the contrary, it may also increase fatigue life by preventing corrosion pitting.

CASTNER PROCESS

The Castner process is a process for manufacturing sodium metal by electrolysis of molten sodium hydroxide at approximately 330 °C. Below that temperature, the melt would solidify; above that temperature, the molten sodium would start to dissolve in the melt.

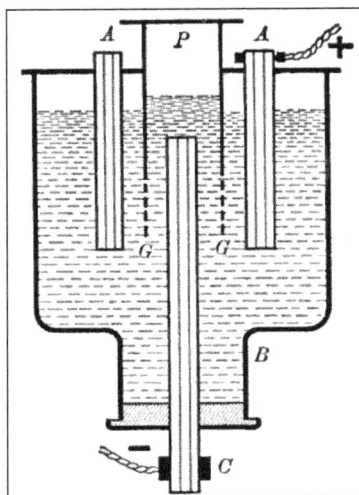

Diagram of Castner process apparatus.

Process Details

The diagram shows a ceramic crucible with a steel cylinder suspended within. Both cathode (C) and anode (A) are made of iron or nickel. The temperature is cooler at the bottom and hotter at the top so that the sodium hydroxide is solid in the neck (B) and liquid in the body of the vessel. Sodium metal forms at the cathode but is less dense than the fused sodium hydroxide electrolyte. Wire gauze (G) confines the sodium metal to accumulating at the top of the collection device (P). The cathode reaction is:

$$2\,Na^+ + 2\,e^- \rightarrow 2\,Na$$

The anode reaction is:

$$2\,OH^- \rightarrow \tfrac{1}{2}\,O_2 + H_2O + 2\,e^-$$

Despite the elevated temperature, some of the water produced remains dissolved in the electrolyte. This water diffuses throughout the electrolyte and results in the reverse reaction taking place on the electrolized sodium metal:

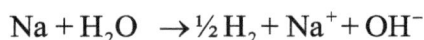

$$Na + H_2O \rightarrow \tfrac{1}{2}\,H_2 + Na^+ + OH^-$$

with the hydrogen gas also accumulating at (P). This, of course, reduces the efficiency of the process.

CASTNER–KELLNER PROCESS

The Castner–Kellner process is a method of electrolysis on an aqueous alkali chloride solution (usually sodium chloride solution) to produce the corresponding alkali hydroxide, invented by American Hamilton Castner and Austrian Karl Kellner in the 1890s.

The first patent for electrolyzing brine was granted in England in 1851 to Charles Watt. His process was not an economically feasible method for producing sodium hydroxide though because it could not prevent the chlorine that formed in the brine solution from reacting with its other constituents. Hamilton Castner solved the mixing problem with the invention of the mercury cell and was granted a U.S. patent in 1892. Austrian chemist, Karl Kellner arrived at a similar solution at about the same time. In order to avoid a legal battle they became partners in 1895, founding the Castner-Kellner Alkali Company, which built plants employing the process throughout Europe. The mercury cell process continues in use to this day. Current-day mercury cell plant operations are criticized for environmental release of mercury leading in some cases to severe mercury poisoning as occurred in Ontario Minamata disease. Due to these concerns, mercury cell plants are being phased out, and a sustained effort is being made to reduce mercury emissions from existing plants.

Process Details

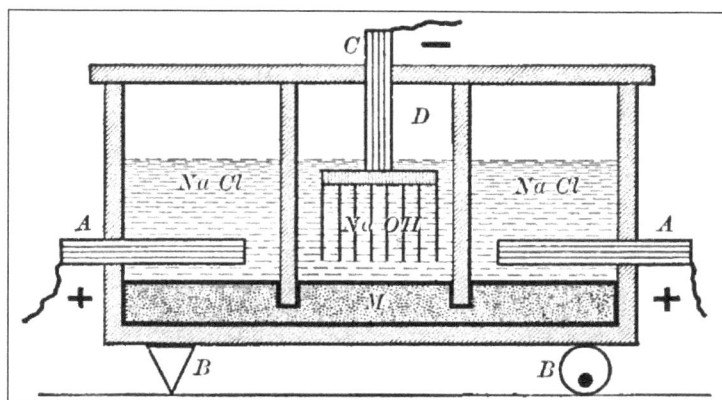

Castner–Kellner apparatus.

The apparatus shown is divided into two types of cells separated by slate walls. The first type, shown on the right and left of the diagram, uses an electrolyte of sodium chloride solution, a graphite anode (A), and a mercury cathode (M). The other type of cell, shown in the center of the diagram, uses an electrolyte of sodium hydroxide solution, a mercury anode (M), and an iron cathode (D). The mercury electrode is common between the two cells. This is achieved by having the walls separating the cells dip below the level of the electrolytes but still allow the mercury to flow beneath them.

The reaction at anode (A) is:

$$Cl^- \rightarrow 1/2\, Cl_2 + e^-$$

The chlorine gas that results vents at the top of the outside cells where it is collected as a byproduct of the process. The reaction at the mercury cathode in the outer cells is,

$$Na^+ + e^- \rightarrow Na\,(amalgam)$$

The sodium metal formed by this reaction dissolves in the mercury to form an amalgam. The mercury conducts the current from the outside cells to the center cell. In addition, a rocking mechanism

(B shown by fulcrum on the left and rotating eccentric on the right) agitates the mercury to transport the dissolved sodium metal from the outside cells to the center cell.

The anode reaction in the center cell takes place at the interface between the mercury and the sodium hydroxide solution.

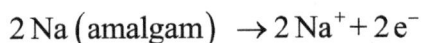

$$2\,Na\,(amalgam) \rightarrow 2\,Na^+ + 2\,e^-$$

Finally at the iron cathode (D) of the center cell the reaction is:

$$2\,H_2O + 2\,e^- \rightarrow 2\,OH^- + H_2$$

The net effect is that the concentration of sodium chloride in the outside cells decreases and the concentration of sodium hydroxide in the center cell increases. As the process continues, some sodium hydroxide solution is withdrawn from center cell as output product and is replaced with water. Sodium chloride is added to the outside cells to replace what has been electrolyzed.

CHLORALKALI PROCESS

The electrolysis of brine is an industrial process for the electrolysis of sodium chloride solutions. It is the technology used to produce chlorine and sodium hydroxide (lye/caustic soda), which are commodity chemicals required by industry. 35 million tons of chlorine were prepared by this process in 1987. Industrial scale production began in 1892.

Old drawing of a chloralkali process plant (Edgewood, Maryland).

Usually the process is conducted on a brine (an aqueous solution of NaCl), in which case NaOH, hydrogen, and chlorine result. When using calcium chloride or potassium chloride, the products contain calcium or potassium instead of sodium. Related processes are known that use molten NaCl to give chlorine and sodium metal or condensed hydrogen chloride to give hydrogen and chlorine.

The process has a high energy consumption, for example over 4 billion kWh per year in West Germany in 1985. Because the process gives equivalent amounts of chlorine and sodium hydroxide (two moles of sodium hydroxide per mole of chlorine), it is necessary to find a use for these products in the same proportion. For every mole of chlorine produced, one mole of hydrogen is produced. Much of this hydrogen is used to produce hydrochloric acid or ammonia, or is used in the hydrogenation of organic compounds.

Cell room of a chlor-alkali plant.

Process Systems

Three production methods are in use. While the mercury cell method produces chlorine-free sodium hydroxide, the use of several tonnes of mercury leads to serious environmental problems. In a normal production cycle a few hundred pounds of mercury per year are emitted, which accumulate in the environment. Additionally, the chlorine and sodium hydroxide produced via the mercury-cell chloralkali process are themselves contaminated with trace amounts of mercury. The membrane and diaphragm method use no mercury, but the sodium hydroxide contains chlorine, which must be removed.

The performance of these devices is governed by the considerations of electrochemical engineering.

Membrane Cell

Basic membrane cell used in the electrolysis of brine. At the anode (A), chloride (Cl^-) is oxidized to chlorine. The ion-selective membrane (B) allows the counterion Na+ to freely flow across, but prevents anions such as hydroxide (OH^-) and chloride from diffusing across. At the cathode (C), water is reduced to hydroxide and hydrogen gas. The net process is the electrolysis of an aqueous solution of NaCl into industrially useful products sodium hydroxide (NaOH) and chlorine gas.

The most common chloralkali process involves the electrolysis of aqueous sodium chloride (a brine) in a membrane cell.

Saturated brine is passed into the first chamber of the cell where the chloride ions are oxidised at the anode, losing electrons to become chlorine gas (A in figure):

$$2\,Cl^- \rightarrow Cl_2 + 2e^-$$

At the cathode, positive hydrogen ions pulled from water molecules are reduced by the electrons provided by the electrolytic current, to hydrogen gas, releasing hydroxide ions into the solution (C in figure):

$$2\,H_2O + 2e^- \rightarrow H_2 + 2\,OH^-$$

The ion-permeable ion exchange membrane at the center of the cell allows the sodium ions (Na^+) to pass to the second chamber where they react with the hydroxide ions to produce caustic soda (NaOH) (B in figure). The overall reaction for the electrolysis of brine is thus:

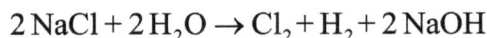

$$2\,NaCl + 2\,H_2O \rightarrow Cl_2 + H_2 + 2\,NaOH$$

A membrane cell is used to prevent the reaction between the chlorine and hydroxide ions. If this reaction were to occur the chlorine would disproportionate to form chloride and hypochlorite ions:

$$Cl_2 + 2\,OH^- \rightarrow Cl^- + ClO^- + H_2O$$

Above about 60 °C, chlorate can be formed:

$$3\,Cl_2 + 6\,OH^- \rightarrow 5\,Cl^- + ClO_3^- + 3\,H_2O$$

Because of the corrosive nature of chlorine production, the anode (where the chlorine is formed) must be made from a non-reactive metal such as titanium, whereas the cathode (where hydroxide forms) can be made from a more easily oxidized metal such as nickel.

Diaphragm Cell

In the diaphragm cell process, there are two compartments separated by a permeable diaphragm, often made of asbestos fibers. Brine is introduced into the anode compartment and flows into the cathode compartment. Similarly to the Membrane Cell, chloride ions are oxidized at the anode to produce chlorine, and at the cathode, water is split into caustic soda and hydrogen. The diaphragm prevents the reaction of the caustic soda with the chlorine. A diluted caustic brine leaves the cell. The caustic soda must usually be concentrated to 50% and the salt removed. This is done using an evaporative process with about three tonnes of steam per tonne of caustic soda. The salt separated from the caustic brine can be used to saturate diluted brine. The chlorine contains oxygen and must often be purified by liquefaction and evaporation.

Mercury Cell

In the mercury-cell process, also known as the Castner–Kellner process, a saturated brine solution floats on top of a thin layer of mercury. The mercury is the cathode, where sodium is produced and forms a sodium-mercury amalgam with the mercury. The amalgam is continuously drawn out of the cell and reacted with water which decomposes the amalgam into sodium hydroxide, hydrogen and mercury. The mercury is recycled into the electrolytic cell. Chlorine is produced at the anode and bubbles out of the cell. Mercury cells are being phased out due to concerns about mercury poisoning from mercury cell pollution such as occurred in Canada and Japan.

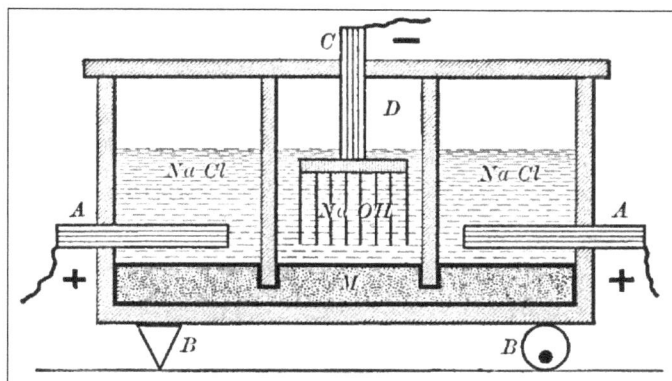

Mercury cell for chloralkali process.

Manufacturer Associations

The interests of chloralkali product manufacturers are represented at regional, national and international levels by associations such as Euro Chlor and The World Chlorine Council.

Laboratory Procedure

Electrolysis can be done with beakers, one containing a brine solution (salt water) and one containing pure water connected by a salt bridge. Anodes are made ideally from platinum group metals, which resist corrosion. Since corrosion is less severe at the cathode, it can be stainless steel or silver.

HALL–HÉROULT PROCESS

The Hall–Héroult process is the major industrial process for smelting aluminium. It involves dissolving aluminium oxide (alumina) (obtained most often from bauxite, aluminium's chief ore, through the Bayer process) in molten cryolite, and electrolysing the molten salt bath, typically in a purpose-built cell. The Hall–Héroult process applied at industrial scale happens at 940–980 °C and produces 99.5–99.8% pure aluminium. Recycled aluminum requires no electrolysis, thus it does not end up in this process.

Process

Challenge

Elemental aluminium cannot be produced by the electrolysis of an aqueous aluminium salt because hydronium ions readily oxidize elemental aluminium. Although a molten aluminium salt could be used instead, aluminium oxide has a melting point of 2072 °C so electrolysing it is im-practical. In the Hall–Héroult process, alumina, Al2O3, is dissolved in molten synthetic cryolite, Na3AlF6, to lower its melting point for easier electrolysis.

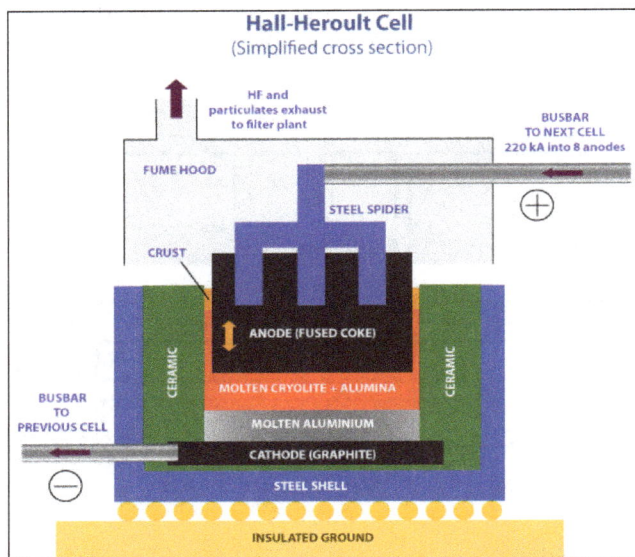

A Hall–Héroult industrial cell.

In the Hall–Héroult process the following simplified reactions take place at the carbon electrodes:

Cathode:

$$Al^{3+} + 3e^- \rightarrow Al$$

Anode:

$$O^{2-} + C \rightarrow CO + 2\,e^-$$

Overall:

$$Al_2O_3 + 3\,C \rightarrow 2\,Al + 3\,CO$$

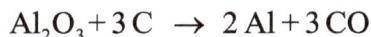

In reality much more CO_2 is formed at the anode than CO:

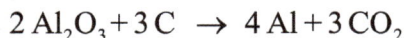

$$2\,Al_2O_3 + 3\,C \rightarrow 4\,Al + 3\,CO_2$$

Pure cryolite has a melting point of 1009 °C ± 1 K. With a small percentage of alumina dissolved in it, its melting point drops to about 1000 °C. Besides having a relatively low melting point, cryolite is used as an electrolyte because among other things it also dissolves alumina well, conducts electricity, dissociates electrolytically at higher voltage than alumina and has a lower density than aluminum at the temperatures required by the electrolysis.

Aluminium fluoride (AlF3) is usually added to the electrolyte. The ratio NaF/AlF$_3$ is called the cryolite ratio and it is 3 in pure cryolite. In industrial production, AlF3 is added so that the cryolite ratio is 2–3 to further reduce the melting point so that the electrolysis can happen at temperatures between 940 and 980 °C. The density of liquid aluminum is 2.3 g/ml at temperatures between 950 and 1000 °C. The density of the electrolyte should be less than 2.1 g/ml so that the molten aluminum separates from the electrolyte and settles properly to the bottom of the electrolysis cell. In addition to AlF3, other additives like lithium fluoride may be added to alter different properties (melting point, density, conductivity etc.) of the electrolyte.

The mixture is electrolysed by passing a low voltage (under 5 V) direct current at 100–300 kA through it. This causes liquid aluminium metal to be deposited at the cathode while the oxygen from the alumina combines with carbon from the anode to produce mostly carbon dioxide.

Cell Operation

Cells in factories are operated 24 hours a day so that the molten material in them will not solidify. Temperature within the cell is maintained via electrical resistance. Oxidation of the carbon anode increases the electrical efficiency at a cost of consuming the carbon electrodes and producing carbon dioxide.

While solid cryolite is denser than solid aluminium at room temperature, liquid aluminium is denser than molten cryolite at temperatures around 1,000 °C (1,830 °F). The aluminium sinks to the bottom of the electrolytic cell, where it is periodically collected. The liquid aluminium is removed from the cell via a siphon every 1 to 3 days in order to avoid having to use extremely high temperature valves and pumps. Alumina is added to the cells as the aluminum is removed. Collected aluminium from different cells in a factory is finally melted together to ensure uniform product and made into e.g. metal sheets. The electrolytic mixture is sprinkled with coke to prevent the anode's oxidation by the oxygen evolved.

The cell produces gases at the anode. The exhaust is primarily CO2 produced from the anode consumption and hydrogen fluoride (HF) from the cryolite and flux (AlF_3). In modern facilities fluorides are almost completely recycled to the cells and therefore used again in the electrolysis. Escaped HF can be neutralized to its sodium salt, sodium fluoride. Particulates are captured using electrostatic or bag filters. The CO_2 is usually vented into the atmosphere.

Agitation of the molten material in the cell increases its production rate at the expense of an increase in cryolite impurities in the product. Properly designed cells can leverage magnetohydrodynamic forces induced by the electrolysing current to agitate the electrolyte. In non-agitating static pool cells the impurities either rise to the top of the metallic aluminium, or else sink to the bottom, leaving high-purity aluminium in the middle area.

Electrodes

Electrodes in cells are mostly coke which has been purified at high temperatures. Pitch resin or tar is used as a binder. The materials most often used in anodes, coke and pitch resin, are mainly residues from petroleum industry and need to be of high enough purity so no impurities end up into the molten aluminum or the electrolyte.

There are two primary anode technologies using the Hall–Héroult process: Söderberg technology and prebake technology.

In cells using Söderberg or self-baking anodes, there is a single anode per electrolysis cell. The anode is contained within a frame and as the bottom of the anode turns mainly into CO_2 during the electrolysis the anode loses mass and being amorphous it slowly sinks within its frame. More material to the top of the anode is continuously added in the form of briquettes made from coke and pitch. The lost heat from the smelting operation is used to bake the briquettes into the carbon form required for the reaction with alumina. The baking process in Söderberg anodes during electrolysis releases more carcinogenic PAHs and other pollutants than electrolysis with prebaked anodes and partially for this reason prebaked anode-using cells have become more common in the aluminium industry. More alumina is added to the electrolyte from the sides of the Söderberg anode after the crust on top of the electrolyte mixture is broken.

Prebake anodes, are baked in very large gas-fired ovens at high temperature before being lowered by various heavy industrial lifting systems into the electrolytic solution. There are usually 24 prebaked anodes in two rows per cell. Each anode is lowered vertically and individually by a computer as the bottom surfaces of the anodes are eaten away during the electrolysis. Compared to Söderberg anodes, computer-controlled prebaked anodes can be brought closer to the molten aluminium layer at the bottom of the cell without any of them touching the layer and interfering with the electrolysis. This smaller distance decreases the resistance caused by the electrolyte mixture and increases the efficiency of prebaked anodes over Söderberg anodes. Prebake technology also has much lower risk of the anode effect, but cells using it are more expensive to build and labor-intensive to use as each prebaked anode in a cell needs to be removed and replaced once it has been used. Alumina is added to the electrolyte from between the anodes in prebake cells.

Prebaked anodes contain a smaller percentage of pitch, as they need to be more solid than Söderberg anodes. The remains of prebaked anodes are used to make more new prebaked anodes. Prebaked anodes are either made in the same factory where electrolysis happens or are brought there from elsewhere.

The inside of the cell's bath is lined with cathode made from coke and pitch. Cathodes also degrade during electrolysis, but much more slowly than anodes do, and thus they need neither be as high in purity nor be maintained as often. Cathodes are typically replaced every 2–6 years. This requires the whole cell to be shut down.

Anode Effect

Anode effect is a situation where too many gas bubbles form at the bottom of the anode and join together forming a layer. This increases the resistance of the cell because smaller areas of the electrolyte touch the anode. These areas of the electrolyte and anode heat up when the density of the electric current of the cell focuses to go through only them. This heats up the gas layer and causes it to expand thus further reducing the surface area where electrolyte and anode are in contact with each other. The anode effect decreases the energy-efficiency and the aluminium production of the cell. It also induces the formation of tetrafluoromethane (CF_4) in significant quantities, increases formation of CO and to a lesser extent also causes the formation of hexafluoroethane (C_2F_6). CF_4 and C_2F_6 are not CFCs, and although not detrimental to the ozone layer, are still potent greenhouse gases. The anode effect is mainly a problem in Söderberg technology cells, not in prebake.

ELECTROLYSIS OF WATER

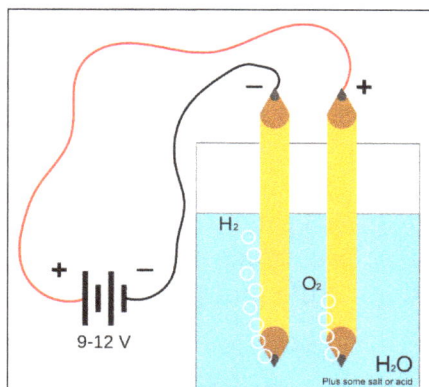

Simple setup for demonstration of electrolysis of water at home.

An AA battery in a glass of tap water with salt showing hydrogen produced at the negative terminal.

Electrolysis of water is the decomposition of water into oxygen and hydrogen gas due to the passage of an electric current.

This technique can be used to make hydrogen gas, a key component of hydrogen fuel, and breathable oxygen gas, or can mix the two into oxyhydrogen - also usable as fuel, though more volatile and dangerous.

It is also called water splitting. It ideally requires a potential difference of 1.23 volts to split water.

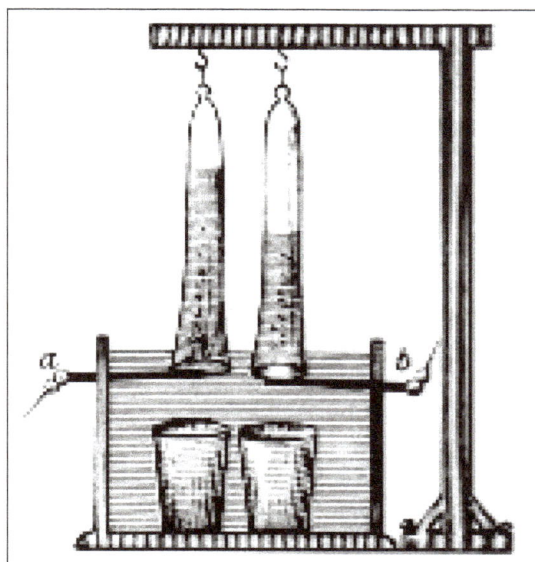

Device invented by Johann Wilhelm Ritter to develop the electrolysis of water.

Jan Rudolph Deiman and Adriaan Paets van Troostwijk used, in 1789, an electrostatic machine to make electricity which was discharged on gold electrodes in a Leyden jar with water. In 1800 Alessandro Volta invented the voltaic pile, and a few weeks later the English scientists William Nicholson and Anthony Carlisle used it for the electrolysis of water. When Zénobe Gramme invented the Gramme machine in 1869 electrolysis of water became a cheap method for the production of hydrogen. A method of industrial synthesis of hydrogen and oxygen through electrolysis was developed by Dmitry Lachinov in 1888.

Principle

A DC electrical power source is connected to two electrodes, or two plates (typically made from some inert metal such as platinum, stainless steel or iridium) which are placed in the water. Hydrogen will appear at the cathode (where electrons enter the water), and oxygen will appear at the anode. Assuming ideal faradaic efficiency, the amount of hydrogen generated is twice the amount of oxygen, and both are proportional to the total electrical charge conducted by the solution. However, in many cells competing side reactions occur, resulting in different products and less than ideal faradaic efficiency.

Electrolysis of pure water requires excess energy in the form of overpotential to overcome various activation barriers. Without the excess energy the electrolysis of pure water occurs very slowly or not at all. This is in part due to the limited self-ionization of water. Pure water has an electrical conductivity about one millionth that of seawater. Many electrolytic cells may also lack the requisite electrocatalysts. The efficiency of electrolysis is increased through the addition of an electrolyte (such as a salt, an acid or a base) and the use of electrocatalysts.

Currently the electrolytic process is rarely used in industrial applications since hydrogen can currently be produced more affordably from fossil fuels.

Equations

$$2\,H_2O \longrightarrow 2\,H_2 \;+\; O_2$$

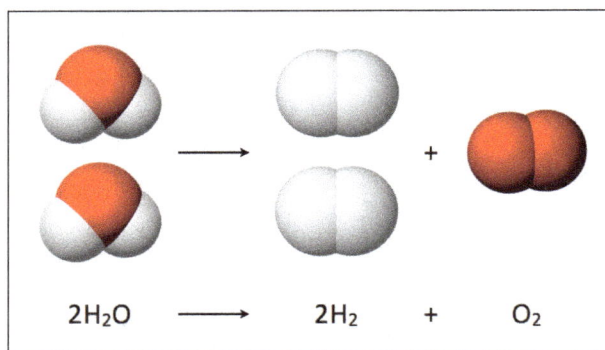

Diagram showing the overall chemical equation.

In pure water at the negatively charged cathode, a reduction reaction takes place, with electrons (e^-) from the cathode being given to hydrogen cations to form hydrogen gas. The half reaction, balanced with acid, is:

Reduction at cathode : $2\,H^+\left(aq\right) + 2\,e^- \rightarrow H_2(g)$

At the positively charged anode, an oxidation reaction occurs, generating oxygen gas and giving electrons to the anode to complete the circuit:

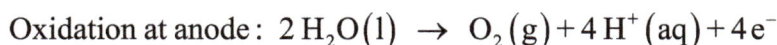

Oxidation at anode : $2\,H_2O(l) \rightarrow O_2(g) + 4\,H^+(aq) + 4\,e^-$

The same half reactions can also be balanced with base as listed below. Not all half reactions must be balanced with acid or base. Many do, like the oxidation or reduction of water listed here. To add half reactions they must both be balanced with either acid or base. The acid-balanced reactions predominate in acidic (low pH) solutions, while the base-balanced reactions predominate in basic (high pH) solutions.

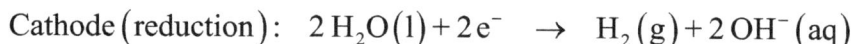

$$\text{Cathode (reduction)}: \quad 2\,H_2O(l) + 2\,e^- \quad \rightarrow \quad H_2(g) + 2\,OH^-(aq)$$
$$\text{Anode (oxidation)}: \quad 2\,OH^-(aq) \quad\quad \rightarrow \quad 1/2\,O_2(g) + H_2O(l) + 2\,e^-$$

Combining either half reaction pair yields the same overall decomposition of water into oxygen and hydrogen:

$$\text{Overall reaction}: \quad 2\,H_2O(l) \quad \rightarrow \quad 2\,H_2(g) + O_2(g)$$

The number of hydrogen molecules produced is thus twice the number of oxygen molecules. Assuming equal temperature and pressure for both gases, the produced hydrogen gas has therefore twice the volume of the produced oxygen gas. The number of electrons pushed through the water is twice the number of generated hydrogen molecules and four times the number of generated oxygen molecules.

Thermodynamics

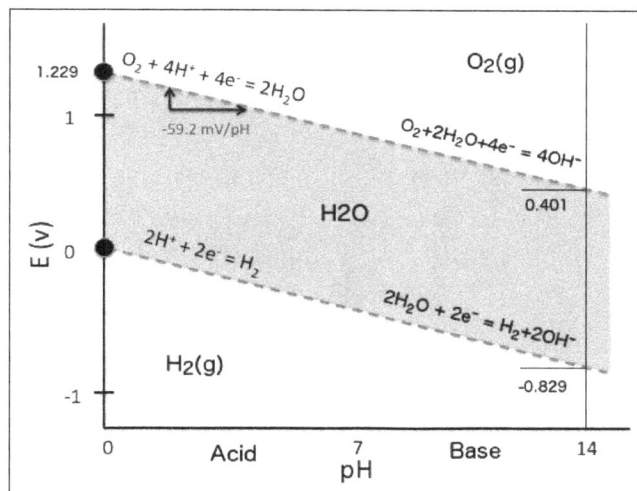

Pourbaix diagram for water, including equilibrium regions for water, oxygen and hydrogen at STP. The vertical scale is the electrode potential of a hydrogen or non-interacting electrode relative to an SHE electrode, the horizontal scale is the pH of the electrolyte (otherwise non-interacting). Neglecting overpotential, above the top line the equilibrium condition is oxygen gas, and oxygen will bubble off of the electrode until equilibrium is reached. Likewise, below the bottom line, the equilibrium condition is hydrogen gas, and hydrogen will bubble off of the electrode until equilibrium is reached.

Decomposition of pure water into hydrogen and oxygen at standard temperature and pressure is not favorable in thermodynamic terms.

Anode (oxidation):	$2\,H_2O(l)$	$\rightarrow$	$O_2(g) + 4\,H^+(aq) + 4e^-$	$E° = +1.23$ V (for the reduction half-equation)
Cathode (reduction):	$2\,H^+(aq) + 2e^-$	$\rightarrow$	$H_2(g)$	$E° = 0.00$ V

Thus, the standard potential of the water electrolysis cell ($E°_{cell} = E°_{cathode} - E°_{anode}$) is −1.23 V at 25 °C at pH 0 ([H^+] = 1.0 M). At 25 °C with pH 7 ([H^+] = 1.0×10^{-7} M), the potential is unchanged based on the Nernst equation. The thermodynamic standard cell potential can be obtained from standard-state free energy calculations to find $\Delta G°$ and then using the equation: $\Delta G° = -nFE°$ (where E° is the cell potential). In practice when an electrochemical cell is "driven" toward completion by applying reasonable potential, it is kinetically controlled. Therefore, activation energy, ion mobility (diffusion) and concentration, wire resistance, surface hindrance including bubble formation (causes electrode area blockage), and entropy, require a greater applied potential to overcome these factors. The amount of increase in potential required is termed the overpotential.

Electrolyte Selection

If the above described processes occur in pure water, H^+ cations will be consumed/reduced at the cathode and OH^- anions will be consumed/oxidised at the anode. This can be verified by adding a pH indicator to the water: the water near the cathode is basic while the water near the anode is acidic. The negative hydroxide ions that approach the anode mostly combine with the positive hydronium ions (H_3O^+) to form water. The positive hydronium ions that approach the cathode mostly combine with negative hydroxide ions to form water. Relatively few hydronium/hydroxide ions reach the cathode/anode. This can cause a concentration overpotential at both electrodes.

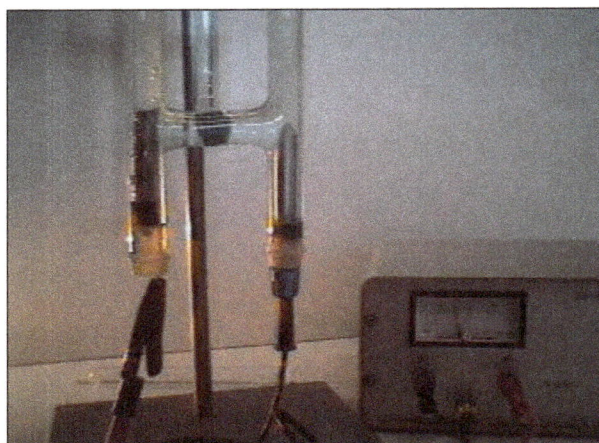

Hoffman voltameter connected to a direct current power supply.

Pure water is a fairly good insulator since it has a low autoionization, $K_w = 1.0×10^{-14}$ at room temperature and thus pure water conducts current poorly, 0.055 µS·cm^{-1}. Unless a very large potential is applied to cause an increase in the autoionization of water the electrolysis of pure water proceeds very slowly limited by the overall conductivity.

If a water-soluble electrolyte is added, the conductivity of the water rises considerably. The electrolyte disassociates into cations and anions; the anions rush towards the anode and neutralize the buildup of positively charged H^+ there; similarly, the cations rush towards the cathode and neutralize the buildup of negatively charged OH^- there. This allows the continuous flow of electricity.

Electrolyte for Water Electrolysis

Care must be taken in choosing an electrolyte, since an anion from the electrolyte is in competition with the hydroxide ions to give up an electron. An electrolyte anion with less standard electrode potential than hydroxide will be oxidized instead of the hydroxide, and no oxygen gas will be produced. A cation with a greater standard electrode potential than a hydrogen ion will be reduced instead, and no hydrogen gas will be produced.

The following cations have lower electrode potential than H^+ and are therefore suitable for use as electrolyte cations: $Li^+, Rb^+, K^+, Cs^+, Ba^{2+}, Sr^{2+}, Ca^{2+}, Na^+$, and Mg^{2+}. Sodium and lithium are frequently used, as they form inexpensive, soluble salts.

If an acid is used as the electrolyte, the cation is H^+, and there is no competitor for the H^+ created by disassociating water. The most commonly used anion is sulfate (SO_4^{2-}), as it is very difficult to oxidize, with the standard potential for oxidation of this ion to the peroxydisulfate ion being +2.010 volts.

Strong acids such as sulfuric acid (H_2SO_4), and strong bases such as potassium hydroxide (KOH), and sodium hydroxide (NaOH) are frequently used as electrolytes due to their strong conducting abilities.

A solid polymer electrolyte can also be used such as Nafion and when applied with a special catalyst on each side of the membrane can efficiently split the water molecule with as little as 1.5 volts. There are also a number of other solid electrolyte systems that have been trialed and developed with a number of electrolysis systems now available commercially that use solid electrolytes.

Pure Water Electrolysis

Electrolyte-free pure water electrolysis has been achieved by using deep-sub-Debye-length nanogap electrochemical cells. When the gap distance between cathode and anode even smaller than Debye-length (1 micron in pure water, around 220 nm in distilled water), the double layer regions from two electrodes can overlap with each other, leading to uniformly high electric field distributed inside the entire gap. Such high electric field can significantly enhance the ion transport inside water (mainly due to migration), further enhancing self-ionization of water and keeping the whole reaction continuing, and showing small resistance between the two electrodes. In this case, the two half-reactions are coupled together and limited by electron-transfer steps (electrolysis current saturated when further reducing the electrode distance).

Techniques

Fundamental Demonstration

Two leads, running from the terminals of a battery, are placed in a cup of water with a quantity of electrolyte to establish conductivity in the solution. Using NaCl (table salt) in an electrolyte solution results in chlorine gas rather than oxygen due to a competing half-reaction. With the correct electrodes and correct electrolyte, such as baking soda (sodium bicarbonate), hydrogen and oxygen gases will stream from the oppositely charged electrodes. Oxygen will collect at the positively charged electrode (anode) and hydrogen will collect at the negatively charged electrode (cathode).

Note that hydrogen is positively charged in the H_2O molecule, so it ends up at the negative electrode. (And vice versa for oxygen).

Note that an aqueous solution of water with chloride ions, when electrolysed, will result in either OH^- if the concentration of Cl^- is low, or in chlorine gas being preferentially discharged if the concentration of Cl^- is greater than 25% by mass in the solution.

Match test used to detect the presence of hydrogen gas.

Hofmann Voltameter

The Hofmann voltameter is often used as a small-scale electrolytic cell. It consists of three joined upright cylinders. The inner cylinder is open at the top to allow the addition of water and the electrolyte. A platinum electrode is placed at the bottom of each of the two side cylinders, connected to the positive and negative terminals of a source of electricity. When current is run through the Hofmann voltameter, gaseous oxygen forms at the anode (positive) and gaseous hydrogen at the cathode (negative). Each gas displaces water and collects at the top of the two outer tubes, where it can be drawn off with a stopcock.

Industrial

Many industrial electrolysis cells are very similar to Hofmann voltameters, with complex platinum plates or honeycombs as electrodes. Generally the only time hydrogen is intentionally produced from electrolysis is for specific point of use application such as is the case with oxyhydrogen torches or when extremely high purity hydrogen or oxygen is desired. The vast majority of hydrogen is produced from hydrocarbons and as a result contains trace amounts of carbon monoxide among other impurities. The carbon monoxide impurity can be detrimental to various systems including many fuel cells.

High-pressure

High-pressure electrolysis is the electrolysis of water with a compressed hydrogen output around 12–20 MPa (120–200 Bar, 1740–2900 psi). By pressurising the hydrogen in the electrolyser, the need for an external hydrogen compressor is eliminated; the average energy consumption for internal compression is around 3%.

High-temperature

High-temperature electrolysis (also HTE or steam electrolysis) is a method currently being investigated for water electrolysis with a heat engine. High temperature electrolysis may be preferable to traditional room-temperature electrolysis because some of the energy is supplied as heat, which is cheaper than electricity, and because the electrolysis reaction is more efficient at higher temperatures.

Nickel/Iron

In 2014, researchers announced an electrolysis system made of inexpensive, abundant nickel and iron rather than precious metal catalysts, such as platinum or iridium. The nickel-metal/nickel-oxide structure is more active than pure nickel metal or pure nickel oxide alone. The catalyst significantly lowers the required voltage. Also nickel–iron batteries are being investigated for use as combined batteries and electrolysis for hydrogen production. Those "battolysers" could be charged and discharged like conventional batteries, and would produce hydrogen when fully charged.

Nanogap Electrochemical Cells

In 2017, researchers reported using nanogap electrochemical cells to achieve high-efficiency electrolyte-free pure water electrolysis at room temperature. In nanogap electrochemical cells, the two electrodes are so close to each other (even smaller than Debye-length in pure water) that the mass transport rate can be even higher than the electron-transfer rate, leading to two half-reactions coupled together and limited by electron-transfer step. Experiments shows that the electrical current density from pure water electrolysis can be even larger than that from 1 mol/L sodium hydroxide solution. The mechanism, "Virtual Breakdown Mechanism", is completely different from the well-established traditional electrochemical theory, due to such nanogap size effect.

Applications

About five percent of hydrogen gas produced worldwide is created by electrolysis. Currently most industrial methods produce hydrogen from natural gas instead, in the steam reforming process. The majority of the hydrogen produced through electrolysis is a side product in the production of chlorine and caustic soda. This is a prime example of a competing side reaction.

$$2\,NaCl + 2\,H_2O \;\rightarrow\; Cl_2 + H_2 + 2\,NaOH$$

The electrolysis of brine, a water/sodium chloride mixture, is only half the electrolysis of water since the chloride ions are oxidized to chlorine rather than water being oxidized to oxygen. Thermodynamically, this would not be expected since the oxidation potential of the chloride ion is less than that of water, but the rate of the chloride reaction is much greater than that of water, causing it to predominate. The hydrogen produced from this process is either burned (converting it back to water), used for the production of specialty chemicals, or various other small-scale applications.

Water electrolysis is also used to generate oxygen for the International Space Station.

Hydrogen may later be used in a fuel cell as a storage method of energy and water.

Efficiency

Industrial Output

Efficiency of modern hydrogen generators is measured by energy consumed per standard volume of hydrogen (MJ/m³), assuming standard temperature and pressure of the H_2. The lower the energy used by a generator, the higher its efficiency would be; a 100%-efficient electrolyser would consume 39.4 kilowatt-hours per kilogram (142 MJ/kg) of hydrogen, 12,749 joules per litre (12.75 MJ/m³). Practical electrolysis (using a rotating electrolyser at 15 bar pressure) may consume 50 kW·h/kg (180 MJ/kg), and a further 15 kW·h (54 MJ) if the hydrogen is compressed for use in hydrogen cars.

Electrolyser vendors provide efficiencies based on enthalpy. To assess the claimed efficiency of an electrolyser it is important to establish how it was defined by the vendor (i.e. what enthalpy value, what current density, etc).

There are two main technologies available on the market, alkaline and proton exchange membrane (PEM) electrolysers. Alkaline electrolysers are cheaper in terms of investment (they generally use nickel catalysts), but less efficient; PEM electrolysers, conversely, are more expensive (they generally use expensive platinum-group metal catalysts) but are more efficient and can operate at higher current densities, and can therefore be possibly cheaper if the hydrogen production is large enough.

Conventional alkaline electrolysis has an efficiency of about 70%. Accounting for the accepted use of the higher heat value (because inefficiency via heat can be redirected back into the system to create the steam required by the catalyst), average working efficiencies for PEM electrolysis are around 80%. This is expected to increase to between 82–86% before 2030. Theoretical efficiency for PEM electrolysers are predicted up to 94%.

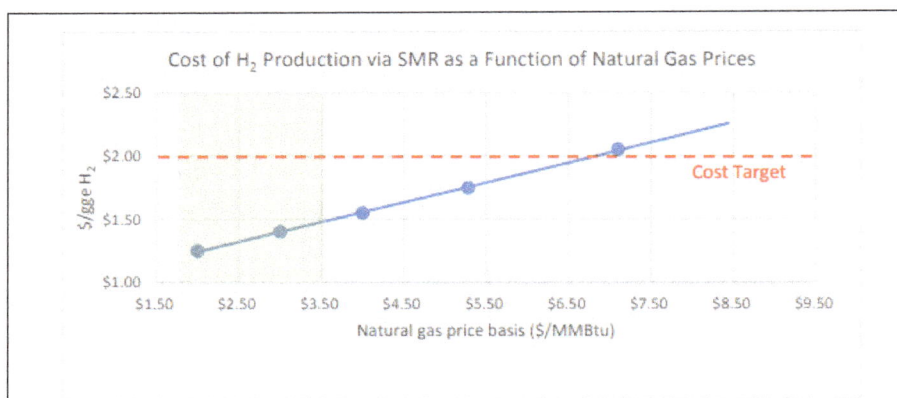

H₂ production cost ($-gge untaxed) at varying natural gas prices.

Considering the industrial production of hydrogen, and using current best processes for water electrolysis (PEM or alkaline electrolysis) which have an effective electrical efficiency of 70–80%, producing 1 kg of hydrogen (which has a specific energy of 143 MJ/kg) requires 50–55 kW·h (180–200 MJ) of electricity. At an electricity cost of $0.06/kW·h, as set out in the Department of Energy hydrogen production targets for 2015, the hydrogen cost is $3/kg. With the range of natural gas prices from 2016 as shown in the graph putting the cost of steam-methane-reformed (SMR) hydrogen at between \$1.20 and \$1.50, the cost price of hydrogen via electrolysis is still over double 2015 DOE hydrogen target prices. The US DOE target price for hydrogen in 2020 is \$2.30/kg, requiring an electricity cost of \$0.037/kW·h, which is achievable given 2018 PPA tenders for wind and solar in many regions. This puts the \$4/gasoline gallon equivalent (gge) H_2 dispensed objective well within reach, and close to a slightly elevated natural gas production cost for SMR.

In other parts of the world, the price of SMR hydrogen is between \$1–3/kg on average. This makes production of hydrogen via electrolysis cost competitive in many regions already, as outlined by Nel Hydrogen and others, including an article by the IEA examining the conditions which could lead to a competitive advantage for electrolysis.

Overpotential

Real water electrolysers require higher voltages for the reaction to proceed. The part that exceeds 1.23 V is called overpotential or overvoltage, and represents any kind of loss and nonideality in the electrochemical process.

For a well designed cell the largest overpotential is the reaction overpotential for the four-electron oxidation of water to oxygen at the anode; electrocatalysts can facilitate this reaction, and platinum alloys are the state of the art for this oxidation. Developing a cheap, effective electrocatalyst for this reaction would be a great advance, and is a topic of current research; there are many approaches, among them a 30-year-old recipe for molybdenum sulfide, graphene quantum dots, carbon nanotubes, perovskite, and nickel/nickel-oxide. Tri-molybdenum phosphide (Mo3P) has been recently found as a promising nonprecious metal and earth-abundant candidate with outstanding catalytic properties that can be used for electrocatalytic processes. The catalytic performance of Mo3P nanoparticles is tested in the hydrogen evolution reaction (HER), indicating an onset potential of as low as 21 mV, H2 formation rate, and exchange current density of 214.7 µmol s−1 g−1 cat (at only 100 mV overpotential) and 279.07 µA cm−2, respectively, which are among the closest values yet observed to platinum. The simpler two-electron reaction to produce hydrogen at the cathode can be electrocatalyzed with almost no overpotential by platinum, or in theory a hydrogenase enzyme. If other, less effective, materials are used for the cathode (e.g. graphite), large overpotentials will appear.

Thermodynamics

The electrolysis of water in standard conditions requires a theoretical minimum of 237 kJ of electrical energy input to dissociate each mole of water, which is the standard Gibbs free energy of formation of water. It also requires energy to overcome the change in entropy of the reaction. Therefore, the process cannot proceed below 286 kJ per mol if no external heat/energy is added.

Since each mole of water requires two moles of electrons, and given that the Faraday constant F represents the charge of a mole of electrons (96485 C/mol), it follows that the minimum voltage necessary for electrolysis is about 1.23 V. If electrolysis is carried out at high temperature, this voltage reduces. This effectively allows the electrolyser to operate at more than 100% electrical efficiency. In electrochemical systems this means that heat must be supplied to the reactor to sustain the reaction. In this way thermal energy can be used for part of the electrolysis energy requirement. In a similar way the required voltage can be reduced (below 1 V) if fuels (such as carbon, alcohol, biomass) are reacted with water (PEM based electrolyzer in low temperature) or oxygen ions (solid oxide electrolyte based electrolyzer in high temperature). This results in some of the fuel's energy being used to "assist" the electrolysis process and can reduce the overall cost of hydrogen produced.

However, observing the entropy component (and other losses), voltages over 1.48 V are required for the reaction to proceed at practical current densities (the thermoneutral voltage).

In the case of water electrolysis, Gibbs free energy represents the minimum *work* necessary for the reaction to proceed, and the reaction enthalpy is the amount of energy (both work and heat) that has to be provided so the reaction products are at the same temperature as the reactant (i.e. standard temperature for the values given above). Potentially, an electrolyser operating at 1.48 V would be 100% efficient.

POLYMER ELECTROLYTE MEMBRANE ELECTROLYSIS

Proton exchange membrane (PEM) electrolysis is the electrolysis of water in a cell equipped with a solid polymer electrolyte (SPE) that is responsible for the conduction of protons, separation of product gases, and electrical insulation of the electrodes. The PEM electrolyzer was introduced to overcome the issues of partial load, low current density, and low pressure operation currently plaguing the alkaline electrolyzer.

However, a recent scientific comparison showed that state-of-the-art alkaline water electrolysis shows competitive or even better efficiencies than PEM water electrolysis. This comparison moreover showed that many of the advantages such as gas purities or high current densities that were ascribed to PEM water electrolysis are also achievable by alkaline water electrolysis. Electrolysis is an important technology for the production of hydrogen to be used as an energy carrier.

With fast dynamic response times, large operational ranges, and high efficiencies, water electrolysis is a promising technology for energy storage coupled with renewable energy sources.

Advantages of PEM Electrolysis

One of the largest advantages to PEM electrolysis is its ability to operate at high current densities. This can result in reduced operational costs, especially for systems coupled with very dynamic energy sources such as wind and solar, where sudden spikes in energy input would otherwise result in uncaptured energy. The polymer electrolyte allows the PEM electrolyzer to operate with a very thin membrane (~100-200 μm) while still allowing high pressures, resulting in low ohmic losses, primarily caused by the conduction of protons across the membrane (0.1 S/cm) and a compressed hydrogen output.

The polymer electrolyte membrane, due to its solid structure, exhibits a low gas crossover rate resulting in very high product gas purity. Maintaining a high gas purity is important for storage safety and for the direct usage in a fuel cell. The safety limits for H_2 in O_2 are at standard conditions 4 mol-% H_2 in O_2.

Science

An electrolyzer is an electrochemical device to convert electricity and water into hydrogen and oxygen, these gases can then be used as a means to store energy for later use. This use can range from electrical grid stabilization from dynamic electrical sources such as wind turbines and solar cells to localized hydrogen production as a fuel for fuel cell vehicles. The PEM electrolyzer utilizes a solid polymer electrolyte (SPE) to conduct protons from the anode to the cathode while insulating the electrodes electrically. Under standard conditions the enthalpy required for the formation of water is 285.9 kJ/mol. A portion of the required energy for a sustained electrolysis reaction is supplied by thermal energy and the remainder is supplied through electrical energy.

Reactions

The actual value for open circuit voltage of an operating electrolyzer will lie between the 1.23 V and 1.48 V depending on how the cell/stack design utilizes the thermal energy inputs. This is however quite difficult to determine or measure because an operating electrolyzer also experiences other voltage losses from internal electrical resistances, proton conductivity, mass transport through the cell and catalyst utilization to name a few.

Anode Reaction

The half reaction taking place on the anode side of a PEM electrolyzer is commonly referred to as the Oxygen Evolution Reaction (OER). Here the liquid water reactant is supplied to catalyst where the supplied water is oxidized to oxygen, protons and electrons.

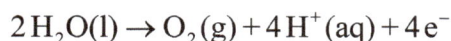

$$2\,H_2O(l) \rightarrow O_2(g) + 4\,H^+(aq) + 4\,e^-$$

Cathode Reaction

The half reaction taking place on the cathode side of a PEM electrolyzer is commonly referred to as the Hydrogen Evolution Reaction (HER). Here the supplied electrons and the protons that have conducted through the membrane are combined to create gaseous hydrogen.

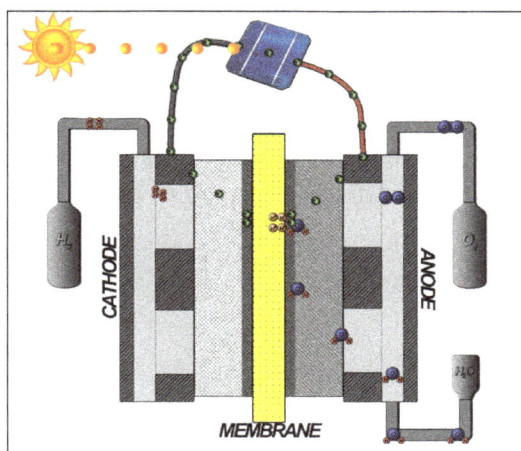

$$4\,H^+(aq) + 4\,e^- \rightarrow 2\,H_2(g)$$

The illustration below depicts a simplification of how PEM electrolysis works, showing the individual half-reactions together along with the complete reaction of a PEM electrolyzer. In this case the electrolyzer is coupled with a solar panel for the production of hydrogen, however the solar panel could be replaced with any source of electricity.

Diagram of PEM electrolyzer cell and the basic principles of operation.

Second Law of Thermodynamics

As per the second law of thermodynamics the enthalpy of the reaction is:

$$\Delta H = \underbrace{\Delta G}_{\text{elec.}} + \underbrace{T \Delta S}_{\text{heat}}$$

Where ΔG is the Gibbs free energy of the reaction, T is the temperature of the reaction and ΔS is the change in entropy of the system.

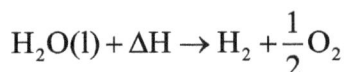

$$H_2O(l) + \Delta H \rightarrow H_2 + \frac{1}{2}O_2$$

The overall cell reaction with thermodynamic energy inputs then becomes:

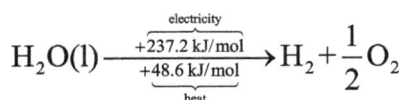

$$H_2O(l) \xrightarrow[\underbrace{+48.6 \text{ kJ/mol}}_{\text{heat}}]{\overset{\text{electricity}}{+237.2 \text{ kJ/mol}}} H_2 + \frac{1}{2}O_2$$

The thermal and electrical inputs shown above represent the minimum amount of energy that can be supplied by electricity in order to obtain an electrolysis reaction. Assuming that the maximum amount of heat energy (48.6 kJ/mol) is supplied to the reaction, the reversible cell voltage V^0_{rev} can be calculated.

Open Circuit Voltage (OCV)

$$V^0_{rev} = \frac{\Delta G^0}{n \cdot F} = \frac{237 \text{ kJ/mol}}{2 \times 96,485 \text{ C/mol}} = 1.23V$$

where n is the number of electrons and F is Faraday's constant. The calculation of cell voltage assuming no irreversibilities exist and all of the thermal energy is utilized by the reaction is referred to as the lower heating value (LHV). The alternative formulation, using the higher heating value (HHV) is calculated assuming that all of the energy to drive the electrolysis reaction is supplied by the electrical component of the required energy which results in a higher reversible cell voltage. When using the HHV the voltage calculation is referred to as the thermoneutral voltage.

$$V^0_{th} = \frac{\Delta H^0}{n \cdot F} = \frac{285.9 \text{ kJ/mol}}{2 \times 96,485 \text{ C/mol}} = 1.48V$$

Voltage Losses

The performance of electrolysis cells, like fuel cells, are typically compared by plotting their polarization curves, which is obtained by plotting the cell voltage against the current density. The primary sources of increased voltage in a PEM electrolyzer (the same also applies for PEM fuel cells) can be categorized into three main areas, Ohmic losses, activation losses and mass transport losses. Due to the reversal of operation between a PEM fuel cell and a PEM electrolyzer, the degree of impact for these various losses is different between the two processes.

$$V_{cell} = E + V_{act} + V_{trans} + V_{ohm}$$

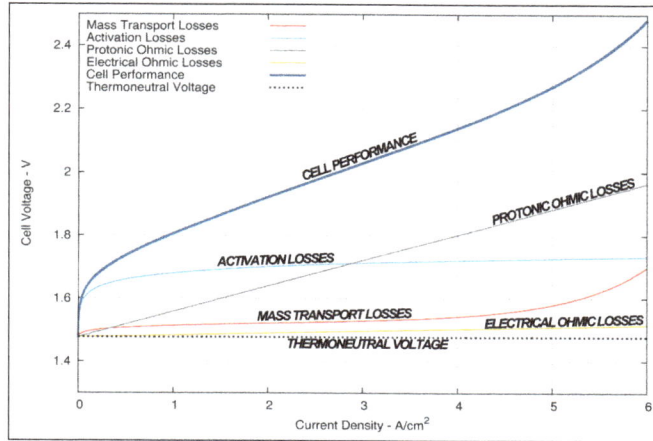

Polarization curve depicting the various losses attributed to PEM electrolysis cell operation.

The performance of a PEM electrolysis system is typically compared by plotting the overpotential versus the cells current density. This essentially results in a curve that represents the power per square centimeter of cell area required to produce hydrogen and oxygen. Conversely to the PEM fuel cell, the better the PEM electrolyzer the lower the cell voltage at a given current density. The figure below is the result of a simulation from the Forschungszentrum Jülich of a 25 cm² single cell PEM electrolyzer under thermoneutral operation depicting the primary sources of voltage loss and their contributions for a range of current densities.

Ohmic Losses

Ohmic losses are an electrical overpotential introduced to the electrolysis process by the internal resistance of the cell components. This loss then requires an additional voltage to maintain the electrolysis reaction, the prediction of this loss follows Ohm's law and holds a linear relationship to the current density of the operating electrolyzer.

$$V = I \cdot R$$

The energy loss due to the electrical resistance is not entirely lost. The voltage drop due to resistivity is associated with the conversion the electrical energy to heat energy through a process known as Joule heating. Much of this heat energy is carried away with the reactant water supply and lost to the environment, however a small portion of this energy is then recaptured as heat energy in the electrolysis process. The amount of heat energy that can be recaptured is dependent on many aspects of system operation and cell design.

$$Q \propto I^2 \cdot R$$

The Ohmic losses due to the conduction of protons contribute to the loss of efficiency which also follows Ohm's law, however without the Joule heating effect. The proton conductivity of the PEM is very dependent on the hydration, temperature, heat treatment, and ionic state of the membrane.

Faradaic Losses and Crossover

Faradaic losses describe the efficiency losses that are correlated to the current, that is supplied without leading to hydrogen at the cathodic gas outlet. The produced hydrogen and oxygen can permeate across the membrane, referred to as crossover. Mixtures of both gases at the electrodes result. At the cathode, oxygen can be catalytically reacted with hydrogen on the platinum surface of the cathodic catalyst. At the anode, hydrogen and oxygen do not react at the iridium oxide catalyst. Thus, safety hazards due to explosive anodic mixtures hydrogen in oxygen can result. The supplied energy for the hydrogen production is lost, when hydrogen is lost due to the reaction with oxygen at the cathode and permeation from the cathode across the membrane to the anode corresponds. Hence, the ratio of the amount of lost and produced hydrogen determines the faradaic losses. At pressurized operation of the electrolyzer the crossover and the correlated faradaic efficiency losses increase.

Hydrogen Compression during Water Electrolysis

Hydrogen evolution due to pressurized electrolysis is comparable to an isothermal compression process, which is in terms of efficiency preferable compared to mechanical isotropical compression. However, the contributions of the afore mentioned faradaic losses increase with operating pressures. Thus, in order to produce compressed hydrogen, the in-situ compression during electrolysis and subsequent compression of the gas have to be pondered under efficiency considerations.

PEM Electrolysis System Operation

PEM high pressure electrolyzer system.

The ability of the PEM electrolyzer to operate, not only under highly dynamic conditions, but also in part-load and overload conditions is one of the reasons for the recently renewed interest in this technology. The demands of an electrical grid are relatively stable and predictable, however when coupling these to energy sources such as wind and solar, the demand of the grid rarely matches the generation of the renewable energy. This means energy produced from renewable sources such as wind and solar must have a buffer, or a means of storing off-peak energy.

PEM Efficiency

When determining the electrical efficiency of PEM electrolysis, the higher heat value (HHV) can be used. This is because the catalyst layer interacts with water as steam. As the process operates at 80 °C for PEM electrolysers the waste heat can be redirected through the system to create the steam, resulting in a higher overall electrical efficiency. The lower heat value (LHV) must be used for alkaline electrolysers as the process within these electrolysers requires water in liquid form and uses alkalinity to facilitate the breaking of the bond holding the hydrogen and oxygen atoms together. The lower heat value must also be used for fuel cells, as steam is the output rather than input.

PEM electrolysis has an electrical efficiency of about 80% in working application, in terms of hydrogen produced per unit of electricity used to drive the reaction. The efficiency of PEM electrolysis is expected to reach 82-86% before 2030, while also maintaining durability as progress in this area continues at a pace.

ELECTROWINNING

Electrowinning, also called electroextraction, is the electrodeposition of metals from their ores that have been put in solution via a process commonly referred to as leaching. Electrorefining uses a similar process to remove impurities from a metal. Both processes use electroplating on a large scale and are important techniques for the economical and straightforward purification of non-ferrous metals. The resulting metals are said to be electrowon.

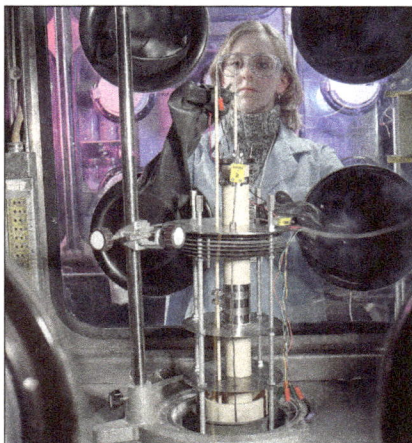

Electrorefining technology converting spent commercial nuclear fuel into metal.

In electrowinning, a current is passed from an inert anode through a liquid *leach* solution containing the metal so that the metal is extracted as it is deposited in an electroplating process onto the cathode. In electrorefining, the anodes consist of unrefined impure metal, and as the current passes through the acidic electrolyte the anodes are corroded into the solution so that the electroplating process deposits refined pure metal onto the cathodes.

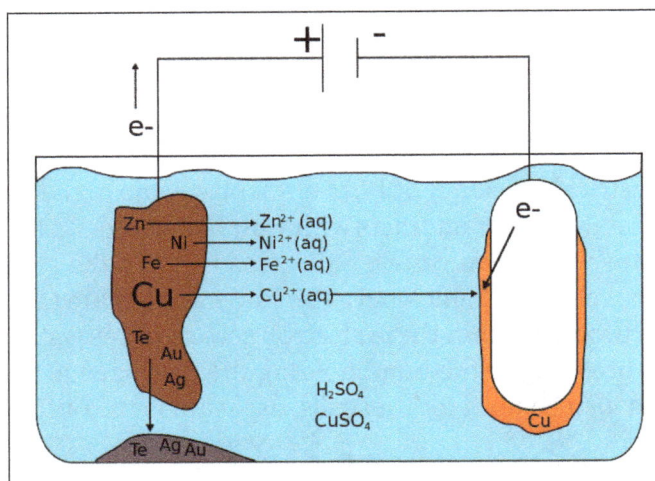

Electrorefining copper.

Electrowinning is the oldest industrial electrolytic process. The English chemist Humphry Davy obtained sodium metal in elemental form for the first time in 1807 by the electrolysis of molten sodium hydroxide.

Electrorefining of copper was first demonstrated experimentally by Maximilian, Duke of Leuchtenberg in 1883.

James Elkington patented the commercial process in 1865 and opened the first successful plant in Pembrey, Wales in 1870. The first commercial plant in the United States was the Balbach and Sons Refining and Smelting Company in Newark, New Jersey in 1883.

Applications

The most common electrowon metals are lead, copper, gold, silver, zinc, aluminium, chromium, cobalt, manganese, and the rare-earth and alkali metals. For aluminium, this is the only production process employed. Several industrially important active metals (which react strongly with water) are produced commercially by electrolysis of their pyrochemical molten salts. Experiments using electrorefining to process spent nuclear fuel have been carried out. Electrorefining may be able to separate heavy metals such as plutonium, caesium, and strontium from the less-toxic bulk of uranium. Many electroextraction systems are also available to remove toxic (and sometimes valuable) metals from industrial waste streams.

Process

Most metals occur in nature in their oxidized form (ores) and thus must be reduced to their metallic forms. The ore is dissolved following some preprocessing in an aqueous electrolyte or in a molten salt and the resulting solution is electrolyzed. The metal is deposited on the cathode (either in solid or in liquid form), while the anodic reaction is usually oxygen evolution. Several metals are naturally present as metal sulfides; these include copper, lead, molybdenum, cadmium, nickel, silver, cobalt, and zinc. In addition, gold and platinum group metals are associated with sulfidic base metal ores. Most metal sulfides or their salts, are electrically conductive and this allows electrochemical redox reactions to efficiently occur in the molten state or in aqueous solutions.

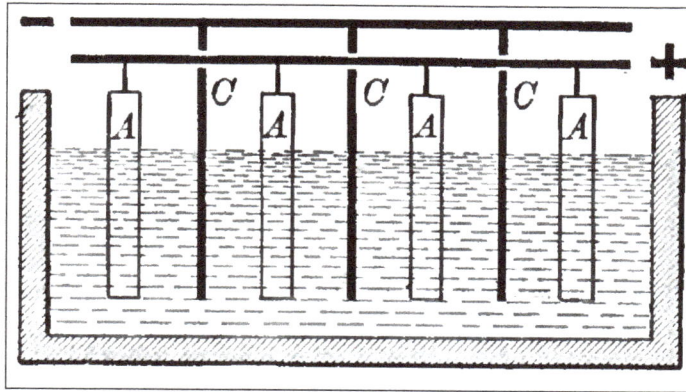

Apparatus for electrolytic refining of copper.

Some metals, such as nickel do not electrolyze out but remain in the electrolyte solution. These are then reduced by chemical reactions to refine the metal. Other metals, which during the processing of the target metal have been reduced but not deposited at the cathode, sink to the bottom of the electrolytic cell, where they form a substance referred to as anode sludge or anode slime. The metals in this sludge can be removed by standard pyrorefining methods.

Because metal deposition rates are related to available surface area, maintaining properly working cathodes is important. Two cathode types exist, flat-plate and reticulated cathodes, each with its own advantages. Flat-plate cathodes can be cleaned and reused, and plated metals recovered. Reticulated cathodes have a much higher deposition rate compared to flat-plate cathodes. However, they are not reusable and must be sent off for recycling. Alternatively, starter cathodes of pre-refined metal can be used, which become an integral part of the finished metal ready for rolling or further processing.

HIGH-TEMPERATURE ELECTROLYSIS

High-temperature electrolysis (also HTE or steam electrolysis) is a technology for producing hydrogen from water at high temperatures.

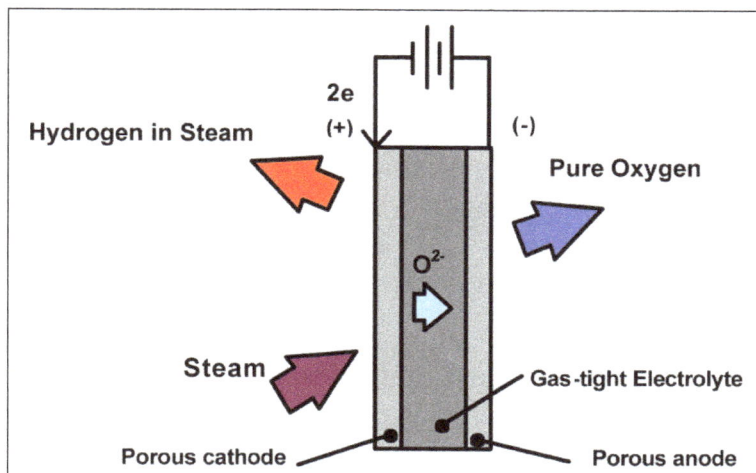

High-temperature electrolysis schema.

Efficiency

High temperature electrolysis is more efficient economically than traditional room-temperature electrolysis because some of the energy is supplied as heat, which is cheaper than electricity, and also because the electrolysis reaction is more efficient at higher temperatures. In fact, at 2500 °C, electrical input is unnecessary because water breaks down to hydrogen and oxygen through thermolysis. Such temperatures are impractical; proposed HTE systems operate between 100 °C and 850 °C.

The efficiency improvement of high-temperature electrolysis is best appreciated by assuming that the electricity used comes from a heat engine, and then considering the amount of heat energy necessary to produce one kg hydrogen (141.86 megajoules), both in the HTE process itself and also in producing the electricity used. At 100 °C, 350 megajoules of thermal energy are required (41% efficient). At 850 °C, 225 megajoules are required (64% efficient).

Materials

The selection of the materials for the electrodes and electrolyte in a solid oxide electrolyser cell is essential. One option being investigated for the process used yttria-stabilized zirconia (YSZ) electrolytes, nickel-cermet steam/hydrogen electrodes, and mixed oxide of lanthanum, strontium and cobalt oxygen electrodes.

Economic Potential

Even with HTE, electrolysis is a fairly inefficient way to store energy. Significant conversion losses of energy occur both in the electrolysis process, and in the conversion of the resulting hydrogen back into power.

At current hydrocarbon prices, HTE can not compete with pyrolysis of hydrocarbons as an economical source of hydrogen.

HTE is of interest as a more efficient route to the production of hydrogen, to be used as a carbon neutral fuel and general energy storage. It may become economical if cheap non-fossil fuel sources of heat (concentrating solar, nuclear, geothermal) can be used in conjunction with non-fossil fuel sources of electricity (such as solar, wind, ocean, nuclear).

Possible supplies of cheap high-temperature heat for HTE are all nonchemical, including nuclear reactors, concentrating solar thermal collectors, and geothermal sources. HTE has been demonstrated in a laboratory at 108 kilojoules (electric) per gram of hydrogen produced, but not at a commercial scale. The first commercial generation IV reactors are expected around 2030.

The Market for Hydrogen Production

Given a cheap, high-temperature heat source, other hydrogen production methods are possible. Thermochemical production might reach higher efficiencies than HTE because no heat engine is required. However, large-scale thermochemical production will require significant advances in materials that can withstand high-temperature, high-pressure, highly corrosive environments.

The market for hydrogen is large (50 million metric tons/year in 2004, worth about $135 billion/ year) and growing at about 10% per year. This market is met by pyrolysis of hydrocarbons to produce the hydrogen, which results in CO2 emissions. The two major consumers are oil refineries and fertilizer plants (each consumes about half of all production). Should hydrogen-powered cars become widespread, their consumption would greatly increase the demand for hydrogen in a hydrogen economy.

Electrolysis and Thermodynamics

During electrolysis, the amount of electrical energy that must be added equals the change in Gibbs free energy of the reaction plus the losses in the system. The losses can (theoretically) be arbitrarily close to zero, so the maximum thermodynamic efficiency of any electrochemical process equals 100%. In practice, the efficiency is given by electrical work achieved divided by the Gibbs free energy change of the reaction.

In most cases, such as room temperature water electrolysis, the electric input is larger than the enthalpy change of the reaction, so some energy is released as waste heat. In the case of electrolysis of steam into hydrogen and oxygen at high temperature, the opposite is true. Heat is absorbed from the surroundings, and the heating value of the produced hydrogen is higher than the electric input. In this case the efficiency relative to electric energy input can be said to be greater than 100%. The maximum theoretical efficiency of a fuel cell is the inverse of that of electrolysis at the same temperature. It is thus impossible to create a perpetual motion machine by combining the two processes.

KOLBE ELECTROLYSIS

The Kolbe electrolysis or Kolbe reaction is an organic reaction named after Hermann Kolbe. The Kolbe reaction is formally a decarboxylative dimerisation of two carboxylic acids (or carboxylate ions) The overall general reaction is:

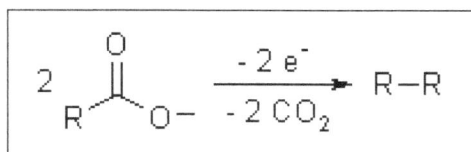

If a mixture of two different carboxylates are used, all combinations of them are generally seen as the organic product structures:

$$3\,R_1COO^- + 3\,R_2COO^- \rightarrow R_1-R_1 + R_1-R_2 + R_2-R_2 + 6\,CO_2 + 6\,e^-$$

The reaction mechanism involves a two-stage radical process: electrochemical decarboxylation gives a radical intermediate, then two such intermediates combine to form a covalent bond. As an example, electrolysis of acetic acid yields ethane and carbon dioxide:

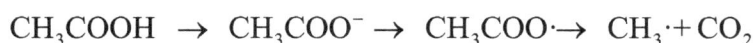

$$CH_3COOH \rightarrow CH_3COO^- \rightarrow CH_3COO\cdot \rightarrow CH_3\cdot + CO_2$$

$$2\,CH_3{\cdot} \rightarrow CH_3CH_3$$

Another example is the synthesis of 2,7-dimethyl-2,7-dinitrooctane from 4-methyl-4-nitrovaleric acid:

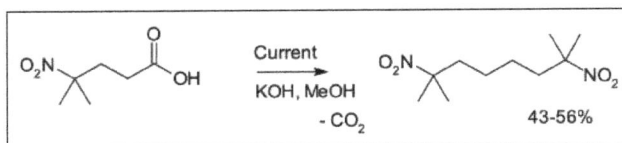

PULSE ELECTROLYSIS

Pulse electrolysis is an alternate electrolysis method that utilises a pulsed direct current to initiate non-spontaneous chemical reactions. Also known as pulsed direct current (PDC) electrolysis, the increased number of variables that it introduces to the electrolysis method can change the application of the current to the electrodes and the resulting outcome. This varies from direct current (DC) electrolysis, which only allows the variation of one value, the voltage applied. By utilising conventional pulse width modulation (PMW), multiple dependent variables can be altered, including the type of waveform, typically a rectangular pulse wave, and the duty cycle, which determines the waveform frequency.

Currently, there has been a focus on theoretical and experimental research into PDC electrolysis in terms of the electrolysis of water to produce hydrogen. Past research has demonstrated that there is a possibility it can result in a higher electrical efficiency in comparison to DC electrolysis. This would allow electrolysis procedures to produce greater volumes of hydrogen with a reduced electrical energy consumption. Although theoretical research has made large promise for the efficiencies and benefits of utilising pulse electrolysis, it has many contradictions including a common issue that it is difficult to replicate the successes of patents experimentally and produces its own negative effects on the electrolyser.

PDC electrolysis is not only confined to the electrolysis of water. Uses in industry such as electroplating and electrocrystallisation are also undergoing research due to the wider range of properties that can be achieved.

The various and alterable effects of using intermittent pulses in PDC electrolysis has resulted in an area of interest that could benefit industry. However, as it is still being researched and has produced conflicting results, a consistent and reliable answer to how dependent electrolysis efficiency is on the properties of an electrical pulse has not been determined, hence, other forms of electrolysis such as polymer electrolyte membrane and alkaline water electrolysis are being used in industry.

With the perspective that the current use of non-renewable fuel sources is a main cause of global environmental problems, hydrogen is being viewed as a possible renewable fuel source replacement. For this to be feasible, the production of hydrogen, through methods such as electrolysis, must be efficient in terms of the energy, cost and time required. Whilst multiple methods of pulse electrolysis have been studied, and experimental results are mixed, the underlying theory behind this experimental approach seems to remain consistent.

Theoretical Concept

When a voltage is applied to an electrolysis cell, immediately following this an Electric Double Layer (EDL), or a diffusion layer, is theoretically formed. This can create a capacitance, or can cause the electrolyser to act as a capacitor. When this is present, excess voltage must be supplied by the direct current to compensate for the loss in the 'capacitor', which rises the required voltage supplied to what is called the thermo-neutral voltage. One of the aims of PDC electrolysis is to overcome this, and theoretically, when the PMW switches the current on, a capacitance will be stored, and when the duty cycle is over, it will be released, continuing the flow of current whilst reducing the EDL that is formed.

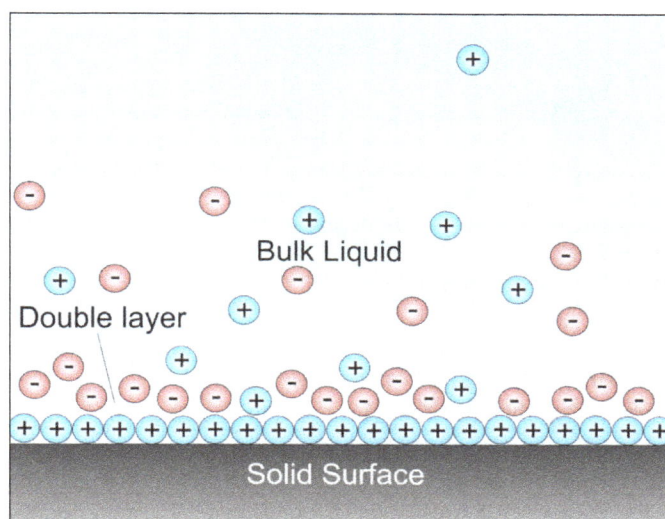

Poláčik and Pospíšil believe that by manipulating the dependent variables, such as the duty cycle, can increase or decrease the effectiveness of pulse electrolysis at reducing this layer. A theoretical equation, the Sand equation, is used to calculate the amount of time required to allow the EDL to fall to zero, and allow PDC electrolysis to achieve its highest efficiencies.

Use in Magnetolisis

Electrolysers require high currents produced by very low voltages. A homopolar generator has the ability to do this, so in Bockris and Ghoroghchian's original experiment in 1985, they followed Faraday's idea. Using a magnetic field of 0.86T produced by permanent magnets, they placed a stainless-steel disc in between. The disc needed a rotation speed of 2000 rpm to reach the correct electrical potential for electrolysis. The difference between Faraday's original model and Bockris and Ghorogchian's is that their disc will only rotate when it is in contact with an electrolyte.

They encountered one large problem, a viscous force created by the electrolyte, that slowed down the motion of the disc. The two ways they could fix this is to rotate the disc and solution together or increase the magnetic field used. The latter being most practicable, the required magnetic field was calculated according to the power consumption rate or producing a cubic meter of hydrogen.

It was discovered a magnetic field of 11T was needed for effective electrolysis, more than 16 times greater than what was originally used. Since superconducting magnets would be required, and they can become too expensive to justify their use, ruling this out as a possible method.

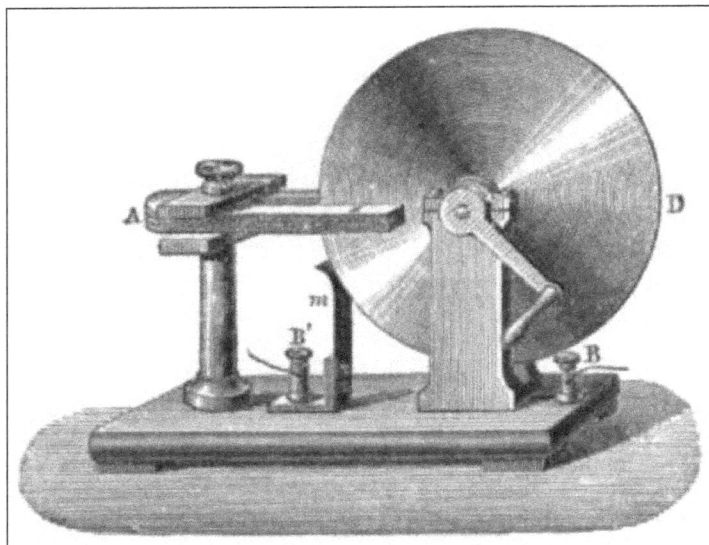

Faraday disk generator the Magnetolysis design was based on.

Their final decision was to use a homopolar generator as an external source of power. This follows Faraday's method more closely.

In this method, a pulse potential was created to take advantage of previous studies that give an effectiveness factor of 2 when either a nickel electrode or a Teflon-bonded platinum electrode was used.

The generator was constructed with a magnetic flux density of 0.6T, a propeller radius of 30 cm and a loop coated with copper strips. To increase the output potential, and reducing the rotation speed required, these were connected in series. Pulses of 2-3V that were sustained for 1ms were achieved.

This was the first instance of a successful application of pulse electrolysis for the production of hydrogen. However, it still presents its own limitations in the possibility for it to be used in industry.

Targeting Resonant Frequency

Mazloomi et al. were first to investigate in 2012 the idea that 'targeting the resonant frequency of the water electrolysis cell' can increase power efficiency, since the main cost of hydrogen production rests with power expenses.

A simplified electrolysis model was used, to which a 'metal oxide semiconductor field effect transistor' (MOSFET) was applied. This filtered the current in the to have a frequency in the range between 0 Hz and 2 MHz. A function generator produced a square wave, duty cycle of 50%, with voltages between -0.5 and +18 V. Throughout the experiment, the frequency was changed within the 20 Hz to 2 MHz range and results were recorded against the cell voltage and current.

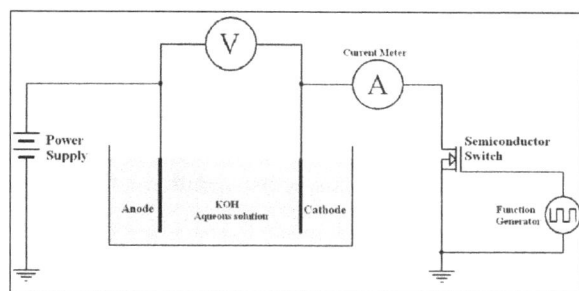

Simplified Experimental Electrolysis Model
used by Mazloomi et al.

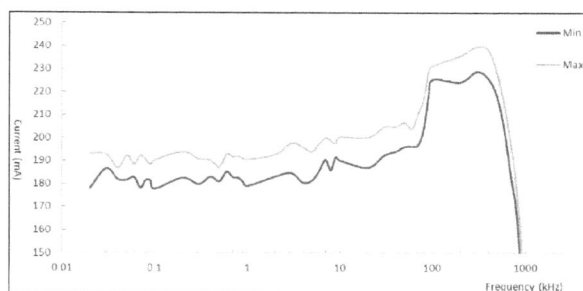

Effect of Changing Frequency in
Water Electrolysis.

An example test showing a positive response is with a 1.5 cm² aluminium electrode, in a 0.1M KOH solution 5 cm apart. The resulting graph shows a natural frequency at 200 kHz, rising the peak current to 230 mA. Assessing the current value achieved in response to the changing frequency showed whether or not the natural frequency was found. Further experimentation found the optimal frequency changed depending on the size of the electrodes, the molarity of the electrolyte and the distance between the electrodes.

It was concluded that reducing 'specific electrical resistance' in an electrolytic cell can lower the required voltage by up to 15% if a high current density and an electrode with a large surface area, is applied.

Conflicting Research

A comparison between a pulsed and non-pulsed dc current electrolysers was explored in 1993 by Shaaban, that demonstrated a non-pulsed current used the least electrical power. This opposes the previous and future works conducted.

The experimental electrolyser separated the anolyte and catholyte compartments and used a 324-Naflon membrane to allow the ion exchange. The distance between the anode, made with platinum coated titanium, and the cathode, stainless steel, was 3mm and was immersed in a 10 weight percent sulfuric acid electrolyte. He conducted tests under several different frequencies that included '0.01 Hz, 0.5 kHz, 5 kHz, i kHz, 10 kHz, 25 kHz, and 40 kHz' and with four duty cycles, '10, 25, 50, and 80%'.

Initial observations revealed that the off-period resulted in a reversal in polarity, causing the reaction to reverse. This effected the cathode, which displayed a 2g loss after experimentation. A diode was input into the circuit to rectify the polarity. However, the cell was prevented from dropping to 0 V during the off-period, maintaining a higher value of 2.3V. This further impacted the experiment, distorting the square wave produced by the function generator Shaaban used, as the electrical potential provided needed to overcome the cell voltage of 2.3V before current could flow. Bokris *et al.* records that current would continue to flow, discharging ions from the EDL, but this was contradicted in this experiment. This only occurred when the diode was in place but it prevented a current spike in the duty cycle as well.

With a 10% duty cycle at a 1 kHz pulse, temperature increases of nearly 7 °C greater than in the non-pulsed experimental electrolysis, were found. Temperature increases can prevent the circuit.

Calculating the power consumption, it was determined a non-pulsed current had power demand losses of 3.5%, and a pulsed current resulted in 13 - 16% losses. This conflicts with Mazloomi *et al.* as the voltage required in pulsed electrolysis will be greater. It also opposes the idea from Bockris *et al.* that the effectiveness of non-pulsed dc current electrolysis increases by a factor of 2 when a pulsed current is applied.

Industrial Uses

The possible increased effect a pulsed current will have on the corrodibility of metals was first looked at by de la Rive in 1837. It was investigated around 60 years later by Coehn regarding the effect of a current with a rectangular waveform, on the plating of zinc deposits, resulting in a successful application for a patent. A full review on using PDC electrolysis in electroplating, also known as electrodeposition or 'pulse plating', was only published in 1954 by Baeyens, this being the first area of research into the use of pulse electrolysis in industry.

A pulsed current can be varied in many ways that increases the possible outcomes and can vary the properties of deposited metals during eletroplating. Hansel and Roy, in their review of the third European Pulse Plating Seminar, concluded that each deposition system must have a unique sequence developed in order to optimise the process and gain the desired results, opposing the inability of traditional plating to be as freely tailored to a situation. The nucleation and crystallisation of the deposition metal is directly affected and can have favourable or unfavourable circumstances if specific conditions are not met. It is reported that pulse plating can encourage nucleation causing grain refinement, and reducing grain size, as well as increasing the deposit density that can improve micro hardness.

These effects were first researched on zinc by Coehn. It was discovered a pulsed current at a high frequency can produce deposits of higher quality, with properties ranging from a smoother finish by the reduction in grain size, as well as lowering its corrosion rate. This is beneficial as it is mainly used as a sacrificial anode in industry.

Advantages

In theoretical electrolysis of water, a voltage of only 1.23 V is required to split water into hydrogen and oxygen, The formation of an EDL increases this to its thermo-neutral voltage of 1.45 V. Minimising the EDL formed during pulse electrolysis is advantageous, as it can reduce the thermo-neutral voltage and the energy input required, increasing energy efficiency.

Disadvantages

Whilst the method of PDC electrolysis has been proven by Ghoroghichian and Bockris in 1952 and 1985 to work extremely well in theory, it is difficult to replicate with consistently positive results in practical experimentation. Hence, the many mechanisms that have been patented are unable to be repeated and used in industry.

According to Shabaan, during the pulse-off period, if the electrolytic cell is not constructed properly, the current polarity can reverse. This can cause the cathode to deteriorate. In electrolysis, the cathode is where the reduction of hydrogen occurs, forming the desired hydrogen gas. Any loss in mass can reduce the speed and effectiveness of the electrolytic reaction, reducing the overall efficiency of the pulse electrolysis method.

Shaaban also states that due to expected internal losses, such as through heat, the current density required will increase, which increases the required voltage. As a result, greater over potentials are needed that further converts to heat.

PATTERSON POWER CELL

The Patterson power cell is an electrolysis device invented by chemist James A. Patterson, which he said created 200 times more energy than it used, and neutralize radioactivity without emitting any harmful radiation. It is one of several cells that some observers classified as cold fusion; cells which were the subject of an intense scientific controversy in 1989, before being discredited in the eyes of mainstream science.

The Patterson power cell is given little credence by scientists. Physicist Robert L. Park describes the device as fringe science in his book Voodoo Science.

Construction

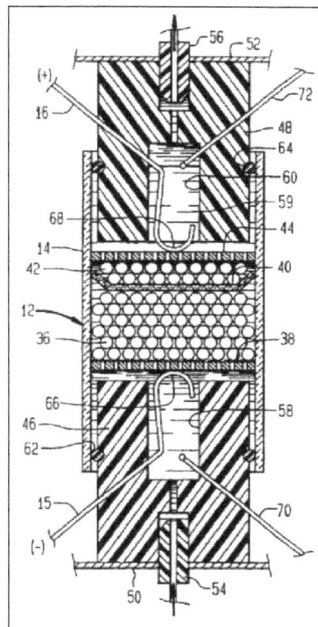

Drawing of the cell.

The cell has a non-conductive housing. The cathode is composed of thousands of sub-millimeter microspheres (co-polymer beads), with a flash coat of copper and multiple layers of electrolytically deposited thin film (650 Angstrom) nickel and palladium. The beads are submerged in water with a lithium sulfate (Li_2SO_4) electrolyte solution. The cell uses lithium sulphate for electrolyte and nano nickel or nickel spheres. It splits water into hydrogen and oxygen.

Company Formed

In 1995, Clean Energy Technologies Inc. was formed to produce and promote the power cell.

Claims and Observations

Patterson variously said it produced a hundred or two hundred times more power than it used. Clean Energy Technologies, Inc. (CETI) representatives promoting the device at the Power-Gen '95 Conference said that an input of 1 watt would generate more than 1,000 watts of excess heat. This supposedly happens as hydrogen or deuterium nuclei fuse together to produce heat through a form of low energy nuclear reaction. The byproducts of nuclear fusion, e.g. a tritium nucleus and a proton or an ^{3}He nucleus and a neutron, have not been detected in a reliable way, leading a vast majority of experts to think that no such fusion is taking place.

It is further claimed that if radioactive isotopes such as uranium are present, the cell enables the hydrogen nuclei to fuse with these isotopes, transforming them into stable elements and thus neutralizing the radioactivity; and this would be achieved without releasing any radiation to the environment and without expending any energy. A televised demonstration on June 11, 1997, on Good Morning America was not conclusive because there was no measurement of the radioactivity of the beads after the test, thus it cannot be discarded that the beads had simply absorbed the uranium ions and become radioactive themselves. In 2002, the neutralization of radioactive isotopes has only been achieved through intense neutron bombardment in a nuclear reactor or large scale high energy particle accelerator, and at a large expense of energy.

When asked about reliability in 1998, Gabe Collins, a chemical engineer at CETI, stated: "When they don't work, it's mostly due to contamination. If you get any sodium in the system it kills the reaction – and since sodium is one of the more abundant elements, it's hard to keep it out."

Patterson has carefully distanced himself from the work of Fleischmann and Pons and from the label of "cold fusion", due to the negative connotations associated to them since 1989. Ultimately, this effort was unsuccessful, and not only did it inherit the label of pathological science, but it managed to make cold fusion look a little more pathological in the public eye. Some cold fusion proponents view the cell as a confirmation of their work, while critics see it as "the fringe of the fringe of cold fusion research", since it attempts to commercialize cold fusion on top of making bad science.

In 2002, John R. Huizenga, professor of nuclear chemistry at the University of Rochester, who was head of a government panel convened in 1989 to investigate the cold fusion claims of Fleischmann and Pons, and who wrote a book about the controversy, said "I would be willing to bet there's nothing to it", when asked about the Patterson Power Cell.

In 2006, Hideo Kozima, professor emeritus of physics at Shizuoka University, has suggested that the byproducts are consistent with cold fusion.

Replications

George H. Miley is a professor of nuclear engineering and a cold fusion researcher who claims to have replicated the Patterson Power Cell. During the 2011 World Green Energy Symposium, Miley stated that his device continuously produces several hundred watts of energy. Earlier results by Miley have not convinced mainstream researchers, who believe that they can be explained by contamination or by misinterpretation of data.

On the television show Good Morning America, Quintin Bowles, professor of mechanical engineering at the University of Missouri–Kansas City, claimed in 1996 to have successfully replicated the Patterson power cell. In the book Voodoo Science, Bowles is quoted as having stated: "It works, we just don't know how it works".

A replication has been attempted at Earthtech, using a CETI supplied kit. They were not able to replicate the excess heat. They looked for cold fusion products, but only found traces of contamination in the electrolyte.

APPLICATIONS OF ELECTROLYSIS

Electrolytic Refining of Metals

The process of electrolytic refining of metals is used to extract impurities from crude metals. Here in this process, a block of crude metal is used as anode, a diluted salt of that metal is used as electrolyte and plates of that pure metal is used as cathode.

Electrolytic Refining of Copper

For understanding the process of electrolytic refining of metals, we will discuss about an example of electrolytic refining of copper. Copper extracted from its ore, known as blister copper, is 98 to 99 % pure but it can easily be made up to 99.95% pure for electrical application by the process of electrorefining.

In this process of electrolysis, we use a block of impure copper as anode or positive electrode, copper sulfate acidified with sulfuric acid, as electrolyte and pure copper plates coated with graphite, as cathode or negative electrode. The copper sulfate splits into positive copper ion (Cu^{++}) and negative sulfate ion (SO_4^{--}). The positive copper ion (Cu^{++}) or cations will move towards negative electrode made of pure copper where it takes electrons from cathode, and becomes Cu atom and is deposited on the graphite surface of the cathode.

On the other hand, the SO_4^{--} will move towards positive electrode or anode where it will receive electrons from anode and become radical SO_4 but as radical SO_4 cannot exist alone, it will attack copper of anode and form $CuSO_4$. This $CuSO_4$ will then dissolve and split in the solution as positive copper ion (Cu^{++}) and negative sulfate ion (SO_4^{--}). These positive copper ions (Cu^{++}) will then move towards negative electrode where it takes electrons from cathode, and become Cu atoms and are deposited on the graphite surface of the cathode. In this way, the copper of impure crude will be transferred and deposited on the graphite surface of the cathode.

The metallic impurities of anode are also merged with SO_4, forming metallic sulfate and dissolve in the electrolyte solution. The impurities like silver and gold, which are not effected by sulfuric acid-copper sulfate solution, will settle down as the anode sludge or mud. At a regular interval of electrolytic refining of copper, the deposited copper is stripped out from the cathode and anode & is replaced by a new block of crude copper.

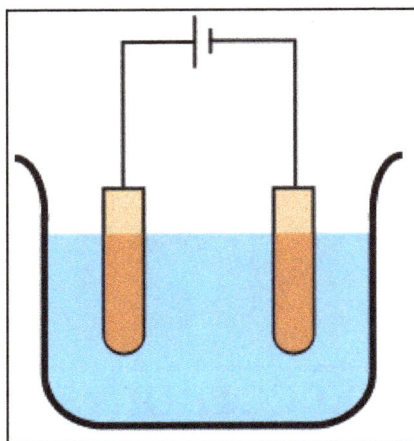

In the process of electrolytic refining of metals or simply electro refining, the cathode is coated by graphite so that the chemical deposited, can be easily stripped off. This is one of the very common applications of electrolysis.

Electroplating

The process of electroplating is theoretically same as electrorefining – only difference is that, in place of graphite coated cathode we have to place an object on which the electroplatinghas to be done. Let's take an example of brass key which is to be copper-platted by using copper electroplating.

Copper Electroplating

We have already stated that copper sulfate splits into positive copper ion (Cu^{++}) and negative sulfate ion (SO_4^{--}) in its solution. For copper electroplating, we use copper sulfate solution as electrolyte, pure copper as anode and an object (a brass key) as cathode. The pure copper rod is connected with positive terminal and the brass key is connected with negative terminal of a battery. While these copper rod and key are immersed into copper-sulfate solution, the copper rod will behave as anode and the key will behave as cathode. As the cathode or the brass key is connected with negative terminal of battery, it will attract the positive cations or Cu^{++} ions and on reaching of Cu^{++} ions on the surface of the brass key, they will receive electrons from it, become neutral copper atom

and are about to be deposited on the surface of the brass key as uniform layer. The sulfate or SO_4^{--} ions move to the anode and extract copper from it into the solution as mentioned in the process of electro-refining. For proper and uniform copper plating, the object (here it is brass key) is being rotated slowly into the solution.

Electroforming

Reproduction of objects by electro-deposition on some sort of mould is known as electroforming.

This is another very useful example among many applications of electrolysis. For that, first we have to take the impression of objects on wax or on other wax like material. The surface of the wax mold which bears exact impression of the object, is coated with graphite powder in order to make it conducting. Then the mold is dipped into the electrolyte solution as cathode. During electrolysis process, the electrolyte metal will be deposited on the graphite coated impressed surface of the mold. After obtaining a layer of desired thickness, the article is removed and the wax is melted to get the reproduced object in form of metal shell. A popular use of electroforming is reproduction of gramophone record dices. The original recording is done on a record of wax composition. This wax mold is then coated with gold powder to make it conducting. Then this mold is dipped into a blue vitriol electrolyte as cathode. The solution is kept saturated by using a copper anode. The copper electroforming on the wax mold produces master plate which is used to stamp a large number of shellac discs.

Electrolysed Water

Electrolysed water (electrolyzed water, EOW, ECA, electrolyzed oxidizing water, electro-activated water or electro-chemically activated water solution) is produced by the electrolysis of ordinary tap water containing dissolved sodium chloride. The electrolysis of such salt solutions produces a solution of hypochlorous acid and sodium hydroxide. The resulting water is a known cleanser and disinfectant/sanitizer.

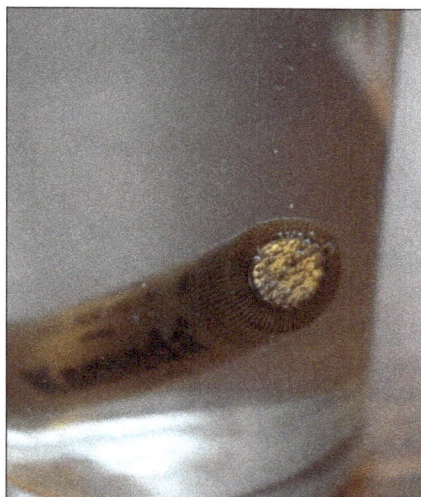

An AA battery in a glass of tap water with salt showing hydrogen produced at the negative terminal.

Creation

The electrolysis occurs in a specially designed reactor which allows the separation of the cathodic and anodic solutions. In this process, hydrogen gas and hydroxide ions can be produced at the cathode, leading to an alkaline solution that consists essentially of sodium hydroxide. At the anode, chloride ions can be oxidized to elemental chlorine, which is present in acidic solution and can be corrosive to metals. If the solution near the anode is acidic then it will contain elemental chlorine, if it is alkaline then it will comprise sodium hydroxide. The key to delivering a powerful sanitising agent is to form hypochlorous acid without elemental chlorine - this occurs at around neutral pH. Hypochlorous is a weak acid and an oxidizing agent. This "acidic electrolyzed water" can be raised in pH by mixing in the desired amount of hydroxide ion solution from the cathode compartment, yielding a solution of Hypochlorous acid (HOCl) and sodium hydroxide (NaOH). A solution whose pH is 7.3 will contain equal concentrations of hypochlorous acid and hypochlorite ion; reducing the pH will shift the balance toward the hypochlorous acid. At a pH between 5.5 and 6.0 approximately 90% of the ions are in the form of hypochlorous acid. In that pH range the disinfectant capability of the solution is more effective than regular sodium hypochlorite (household bleach).

Efficient Disinfectant

Both sodium hydroxide and hypochlorous acid are efficient disinfecting agents; as mentioned above, the key to effective sanitation is to have a high proportion of hypochlorous acid present, this happens between acidic and neutral pH conditions.

EOW will kill spores and many viruses and bacteria.

Electrolysis units sold for industrial and institutional disinfectant use and for municipal water-treatment are known as chlorine generators. These avoid the need to ship and store chlorine, as well as the weight penalty of shipping prepared chlorine solutions. In March, 2016 inexpensive units have become available for home or small business users.

EPA Registration

Although the field of electro-chemical activation ("ECA") technology has existed for more than 40 years, companies producing such solutions have only recently approached the U.S. Environmental Protection Agency (EPA) seeking registration. Recently, a number of companies that manufacture electrolytic devices have sought and received EPA registration as a disinfectant.

Drawbacks

Electrolyzed water loses its potency fairly quickly, so it cannot be stored for long. Electrolysis machines can be but are not necessarily expensive. In some but not all instances the electrolysis process needs to be monitored frequently for the correct potency.

Sodium Hydroxide

Sodium hydroxide, NaOH, also known as lye and caustic soda, is one of the most important of all industrial chemicals. It is produced at the rate of 25 billion pounds a year in the United States alone. The major method for producing it is the electrolysis of brine or "salt water," a solution of common salt, sodium chloride in water. Chlorine and hydrogen gases are produced as valuable byproducts.

When an electric current is passed through salt water, the negative chloride ions, Cl^-, migrate to the positive anode and lose their electrons to become chlorine gas.

(The chlorine atoms then pair up to form Cl_2 molecules.) Meanwhile, sodium ions, Na^+, are drawn to the negative cathode. But they do not pick up electrons to become sodium metal atoms as they do in molten salt, because in a water solution the water molecules themselves pick up electrons more easily than sodium ions do.

The hydroxide ions, together with the sodium ions that are already in the solution, constitute sodium hydroxide, which can be recovered by evaporation.

This so-called chloralkali process is the basis of an industry that has existed for well over a hundred years. By electricity, it converts cheap salt into valuable chlorine, hydrogen and sodium hydroxide. Among other uses, the chlorine is used in the purification of water, the hydrogen is used in the hydrogenation of oils, and the lye is used in making soap and paper.

References

- Electrolysis, chemistry-general, chemistry, science-and-technology: encyclopedia.com, Retrieved 30 April, 2019

- Kutz, Myer (2005-06-02). "Protective coatings for aluminum alloys". Handbook of Environmental Degradation of Materials. Norwich, NY: William Andrew. P. 353. ISBN 978-0-8155-1749-8

- Faraday, topicreview, genchem: purdue.edu, Retrieved 13 March, 2019

- "What are the types of Anodising and which materials can you Anodise?". Www.manufacturingnetwork.com. Archived from the original on 2015-11-26. Retrieved 2015-11-25

- Faradays-first-and-second-laws-of-electrolysis: electrical4u.com, Retrieved 29 June, 2019

- "Anodizing and the environment". Archived from the original on 8 September 2008. Retrieved 2008-09-08

- Applications-of-electrolysis-electroplating-electroforming-electrorefining: electrical4u.com, Retrieved 30 January, 2019

- Huang, Yu-Ru; Yen-Con Hung; Shun-Yao Hsu; Yao-Wen Huang; Deng-Fwu Hwang (April 2008). "Application of electrolyzed water in the food industry". Food Control. 19 (4): 329–345. Doi:10.1016/j.foodcont.2007.08.012. ISSN 0956-7135

- Electrolysis-Production-sodium-hydroxide-chlorine-hydrogenm, 2351, pages: jrank.org,

Electroanalytical Methods

The techniques in analytical chemistry that deal with the study of an analyte is known as electroanalytical method. It measures the potential and current present in an electrochemical cell. The main electroanalytical method are coulometry and voltammetry. The topics elaborated in this chapter will help in gaining a better perspective about electroanalytical method.

Electroanalytical methods, which present a prior accumulation step under convective mass transport of the target analytes to the electrode followed by a measuring step at which the target species are 'stripped away' from the electrode generating a current magnitude equivalent to their concentration, are known as stripping analytical methods. This two-step technique also named 'stripping analysis' can be performed by applying different manners of deposition and stripping of the analytical element. The works referenced during this text used predominantly anodic stripping voltammetry (ASV), stripping chronopotentiometry (SCP), and potentiometric stripping analysis (PSA). The difference among those techniques is related to the stripping step, more specifically to the manner that the electroactive species are stripped from the electrode and to the correspondent registered signal. Stripping voltammetry consists in applying a potential (linear or pulsed scan) to the electrode whilst the current is registered; the direction of the stripping potential distinguishes between anodic or cathodic stripping voltammetry. On the other hand, potentiometric stripping methods measure the potential variation during the stripping process, which can be performed either by applying a constant current (cathodic or anodic SCP) or by the action of an oxidant agent in solution such as dissolved oxygen (PSA). Potentiometric stripping techniques are less susceptive to organic compounds generally present in solution than voltammetric ones, which can be tremendously important for the analysis of complex samples such as biological fluids with minimal sample treatment (simple sample dilution).

After the discovery of preconcentrating electroactive species on the mercury electrode drop, polarography ascended in popularity for analytical purposes attributable to the huge increase of sensitivity attained by accumulation step. This preconcentration advent was extended to other electrodes allowing increments of sensitivity of 100–1000 times and improvement of the detection limits (concentrations lower than 10– 10 mol l– 1).9 Additional to the remarkable sensitivity, advantages such as low-cost, simple and portable equipment for on-site analysis, rapid and reliable methods, as well as the easy miniaturization of electroanalytical systems put in evidence electroanalysis in many fields of research.

Mercury film coated on glassy carbon electrodes (GCEs) appeared as an attractive alternative to mercury drop electrodes for metal determination employing stripping analysis. Around 30 elements of the periodic table can be determined by employing stripping analysis at mercury electrodes.Complementary to mercury electrodes, bare gold electrodes are the most suitable sensors for the determination of mercury, arsenic, selenium, and copper by using stripping analysis. Carbon paste electrodes (CPEs, manufactured by mixing carbon powder, agglutinating oil, and a chemical modifier agent) represent another type of electrode applied in electroanalysis since their discovery in 1960s. The typical chemical modifier is a ligand which links specifically to the target species (similar strategy adopted in the adsorptive stripping analysis where the chemical modifier molecules surround the mercury drop). The simplicity of construction and the possibility of renewing the electrode surface by simple mechanical polishing are the main positive points of CPEs; nevertheless, the same mechanical polishing causes lack of reproducibility between measurements. An additional type of electrode widely used for analytical applications is the screen-printed electrode (SPE). Printer machines can produce large amounts of SPEs using ceramic plates as substrates. The common three electrode system (working, counter, and reference electrodes) can be printed at the same substrate by applying commercial-available inks (carbon ink for working and counter electrodes and silver chloride ink for the reference one). Chemical modifiers can also be incorporated in the carbon ink in order to produce electrodes with improved selectivity and sensitivity. SPEs have found massive importance as source of disposable sensors for portable devices such as the glucose sensor for home-diagnostic of diabetes patients. Additionally, SPEs were successfully applied for stripping analysis determination of heavy metals.

The more recent interest of electrochemists and analytical electrochemists is the development of new materials, especially carbon based materials such as pyrolytic graphite, boron-doped diamond, and carbon nanotubes. Low potential detection (more selective), increase of sensitivity (lower detection limits), and widening of the window potential range (investigation of very high potential processes not before studied) are examples of promising features obtained by exploiting those new materials.

New sources for sensor development especially attending the green chemistry concept are also a demanding direction of research by analytical electrochemists. The elimination of toxic materials and reagents from standard analytical procedures such as the highly toxic mercury has recently received elegant solutions. An example is the bismuth film electrode which presented excellent performance for the electrochemical stripping analysis determination of heavy metals with similar performance of mercury electrodes.

COULOMETRY

This method is the measurement of the quantity of the electricity. This is mainly estimated by the reaction of electrode. There are mainly two types of coulometric techniques. They are as follows:

- Controlled potential coulometry.

- Constant current coulometry.

Principle

The main principle involved in the coulometry is the measurement of the quantity of the electricity which is directly proportional to the chemical reaction at the electrode. This is given by the Faraday's first law:

$$W = \frac{M_r \times Q}{96,487n}$$

where Q is the consumed current; M_r is the relative molecular weight.

The coulometric methods are mainly based on the measurement of the quantity of the electricity. The sample which is to be determined undergoes the reaction at the electrode which is measured at the electrode. The completion of the reaction is indicated by the decrease in the current to zero. This can be measured by the coulometer. The substance which is to be determined is first electrolyzed by the constant current. Then the total current is determined by the following equation:

Total current = product current × time

In electrolysis, in the controlled potential coulometry, the quantity of the current Q is given by the following equation:

$$Q = \int_0^\tau I_t \, dt$$

where I_t is the current at time t.

Then in the concentration terms it is given by the following equation:

$$C_t = C_o e^{-kt}$$

where C_t is the concentration of the electrolyte at time t; C_o is the initial concentration.

Instrumentation

In the instrumentation of the coulometry, mainly two types of electrodes are used: one is the reference electrode and another is the working electrode.

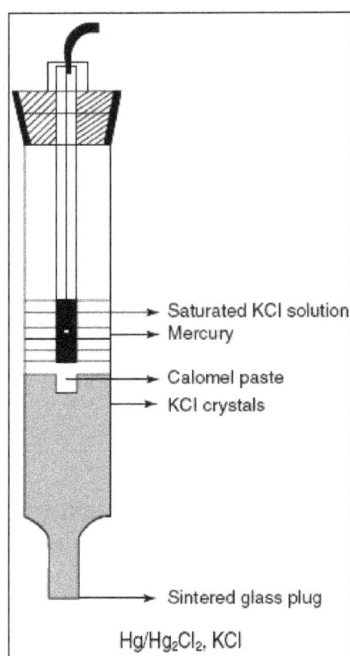

Reference electrode.

Generally saturated calomel electrode is used as the reference electrode. It consists of porous disc at the base of the electrode which is clogged. Above it, the glass tube is filled with the potassium chloride crystals. And above that it is filled with the calomel paste which is prepared by grinding of mercury chloride with pure mercury and minute millilitre of the saturated potassium chloride solution. Then pure mercury is placed in the electrode vessel. The advantages are the following: the easy to construct and highly stable.

The reaction is the following:

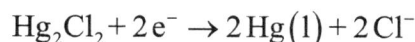

$$Hg_2Cl_2 + 2e^- \rightarrow 2Hg(l) + 2Cl^-$$

Platinum hydrogen electrode is used as the working electrode.

The apparatus used in the coulometry is as follows:

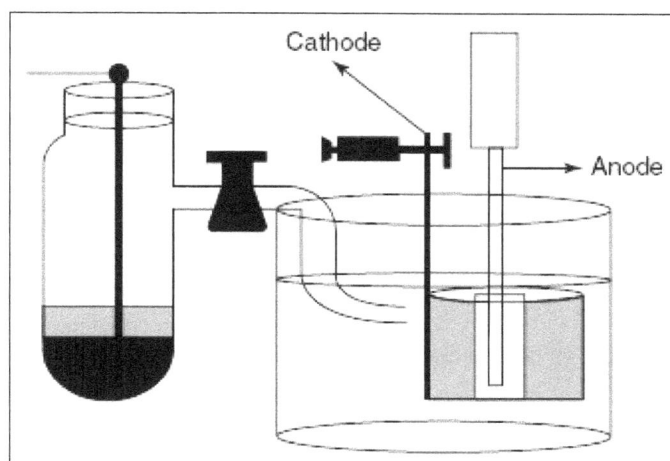

Working electrode.

It consists of the working electrode and the reference electrode and these electrodes are connected to the coulometer. The measured reading is plotted on the graph as follows:

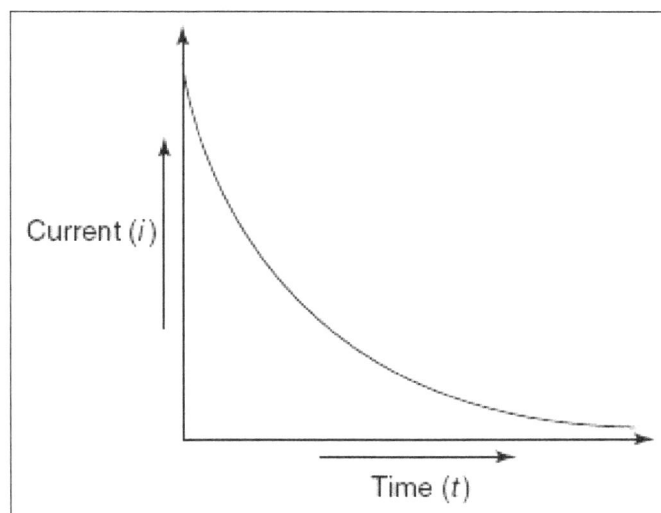

Coulometry titration curve.

Two electrodes are immersed in the sample solution which can be measured. Then the constant potential or constant current is passed through the electrode. Next, the chemical reaction takes place at the working electrode and is compared with that of the reference electrode. The completion of the reaction is indicated by the decrease of the current which is measured by the coulometer.

Coulometric Titrations

Constant coulometric method is commonly known as the coulometric titration. In coulometric titrations, the reagent is generated electrically and determined by the current and by the time. It should be of 100% efficiency and the reagent generated should react with the sample solution. The main principle involved in the coulometric titration is the generation of the titrant by electrolysis. Then a large amount of titrant solution is added to the sample solution. Then the sample solution is electrolysed at the anode surface. As the electrolysis proceeds, the anode potential is increased. Then the addition of the titrant solution decreases the potential by decreasing the current. The end point is determined by the any of the end point detection method.

Example: The sample solutions containing the ferrous ions are added to the excess amount of the Ce (III) ion solution.

$$Fe^{+2} \rightarrow Fe^{+3} + e^-$$

$$Ce^{+4} + Fe^{+2} \rightarrow Ce^{+3} + Fe^{+3}$$

The following are the advantages of the coulometric titrations:

- Standard solutions are not required.

- Reagent is generated.

- No need of the dilution of the sample solution.

- The method is readily adopted than other methods.

The following are the limitations of the coulometric titrations:

- Generation is difficult.

- Inferences are more.

The detection of the end points in the coulometric titrations is done by the following:

- By the chemical indicators: These are added to the sample solution. The only requirement for these reagents should be electroinactive.

- Examples:

  ◦ Methyl orange.

  ◦ Dichlorofluorescein.

  ◦ Eosin.

- Potentiometric end point detection method: When the pair of electrodes are placed in the sample solution it shows the potential difference by the addition of the titrant or by the change in the concentration of the ions. To measure the electromotive force of the electrode, system is measured by the potentiometer or by the electronic voltameter.

- Amperometric method: This method is mainly based on the current produced which is directly proportional to the concentration of the electroactive substance.

- By the spectrophotometric method.

Applications

- Used in the determination of the thickness of the metallic coatings.

- Used in the determination of the total anti-oxidant capacity of the anti-oxidants.

- Used in the determination of the total carbon in ferrous and non-ferrous metals.

- Used in the determination of the picric acid.

- Used in the separation of the nickel and cobalt.

- Used in the analysis of the radioactive materials.

- Used in the determination n-values of the organic compounds.

- Used in the determination of the environment pollutants.

VOLTAMMETRY

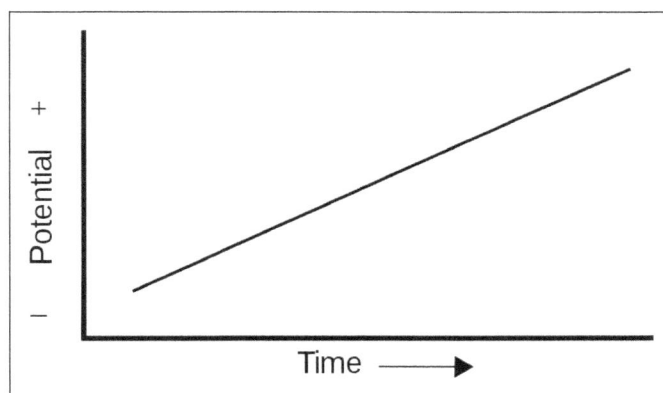

Linear potential sweep.

Voltammetry is a category of electroanalytical methods used in analytical chemistry and various industrial processes. In voltammetry, information about an analyte is obtained by measuring the current as the potential is varied. The analytical data for a voltammetric experiment comes in the form of a voltammagram which plots the current produced by the analyte versus the potential of the working electrode.

Three Electrode System

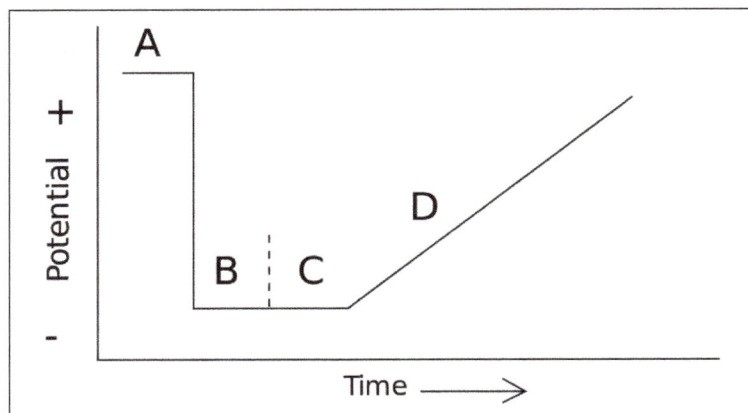

Potential as a function of time for anodic stripping voltammetry.

Voltammetry experiments investigate the half-cell reactivity of an analyte. Voltammetry is the study of current as a function of applied potential. These curves I = f(E) are called voltammograms. The potential is varied arbitrarily either step by step or continuously, and the actual current value is measured as the dependent variable. The opposite, i.e., amperometry, is also possible but not common. The shape of the curves depends on the speed of potential variation (nature of driving force) and on whether the solution is stirred or quiescent (mass transfer). Most experiments control the potential (volts) of an electrode in contact with the analyte while measuring the resulting current (amperes).

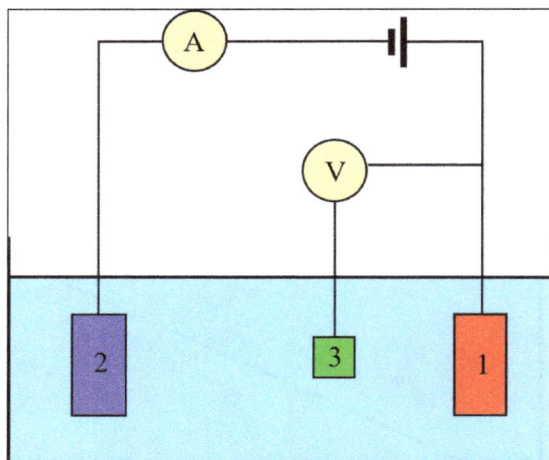

Three-electrode setup: (1) working electrode; (2) counter electrode; (3) reference electrode.

To conduct such an experiment one requires at least two electrodes. The working electrode, which makes contact with the analyte, must apply the desired potential in a controlled way and facilitate the transfer of charge to and from the analyte. A second electrode acts as the other half of the cell. This second electrode must have a known potential with which to gauge the potential of the working electrode, furthermore it must balance the charge added or removed by the working electrode. While this is a viable setup, it has a number of shortcomings. Most significantly, it is extremely difficult for an electrode to maintain a constant potential while passing current to counter redox events at the working electrode.

To solve this problem, the roles of supplying electrons and providing a reference potential are divided between two separate electrodes. The reference electrode is a half cell with a known reduction potential. Its only role is to act as reference in measuring and controlling the working electrode's potential and at no point does it pass any current. The auxiliary electrode passes all the current needed to balance the current observed at the working electrode. To achieve this current, the auxiliary will often swing to extreme potentials at the edges of the solvent window, where it oxidizes or reduces the solvent or supporting electrolyte. These electrodes, the working, reference, and auxiliary make up the modern three electrode system.

There are many systems which have more electrodes, but their design principles are generally the same as the three electrode system. For example, the rotating ring-disk electrode has two distinct and separate working electrodes, a disk and a ring, which can be used to scan or hold potentials independently of each other. Both of these electrodes are balanced by a single reference and auxiliary combination for an overall four electrode design. More complicated experiments may add working electrodes as required and at times reference or auxiliary electrodes.

In practice it can be important to have a working electrode with known dimensions and surface characteristics. As a result, it is common to clean and polish working electrodes regularly. The auxiliary electrode can be almost anything as long as it doesn't react with the bulk of the analyte solution and conducts well. It is (or was?) common to use mercury as working electrode e.g. DME and HMDE, and also as auxiliary, and the voltammetry method is then known as polarography. The reference is the most complex of the three electrodes; there are a variety of standards used and it is worth investigating elsewhere. For non-aqueous work, IUPAC recommends the use of the ferrocene/ferrocenium couple as an internal standard. In most voltammetry experiments, a bulk electrolyte (also known as a supporting electrolyte) is used to minimize solution resistance. It is possible to run an experiment without a bulk electrolyte, but the added resistance greatly reduces the accuracy of the results. With room temperature ionic liquids, the solvent can act as the electrolyte.

Theory

Data analysis requires the consideration of kinetics in addition to thermodynamics, due to the temporal component of voltammetry. Idealized theoretical electrochemical thermodynamic relationships such as the Nernst equation are modeled without a time component. While these models are insufficient alone to describe the dynamic aspects of voltammetry, models like the Tafel equation and Butler–Volmer equation lay the groundwork for the modified voltammetry relationships that relate theory to observed results.

Types of voltammetry

- Linear sweep voltammetry.

- Staircase voltammetry.

- Squarewave voltammetry.

- Cyclic voltammetry: A voltammetric method that can be used to determine diffusion coefficients and half cell reduction potentials.

- Anodic stripping voltammetry: A quantitative, analytical method for trace analysis of metal cations. The analyte is deposited (electroplated) onto the working electrode during a deposition step, and then oxidized during the stripping step. The current is measured during the stripping step.

- Cathodic stripping voltammetry: A quantitative, analytical method for trace analysis of anions. A positive potential is applied, oxidizing the mercury electrode and forming insoluble precipitates of the anions. A negative potential then reduces (strips) the deposited film into solution.

- Adsorptive stripping voltammetry: A quantitative, analytical method for trace analysis. The analyte is deposited simply by adsorption on the electrode surface (i.e., no electrolysis), then electrolyzed to give the analytical signal. Chemically modified electrodes are often used.

- Alternating current voltammetry.

- Polarography: A subclass of voltammetry where the working electrode is a dropping mercury electrode (DME), useful for its wide cathodic range and renewable surface.

- Rotated electrode voltammetry: A hydrodynamic technique in which the working electrode, usually a rotating disk electrode (RDE) or rotating ring-disk electrode (RRDE), is rotated at a very high rate. This technique is useful for studying the kinetics and electrochemical reaction mechanism for a half reaction.

- Normal pulse voltammetry.

- Differential pulse voltammetry.

- Chronoamperometry.

Applications

Voltammetric Sensors

A number of voltammetric systems are produced commercially for the determination of specific species that are of interest in industry and research. These devices are sometimes called electrodes but are, in fact, complete voltammetric cells and are better referred to as sensors. These sensors can be employed for the analysis of various organic and inorganic analytes in various matrices.

The Oxygen Electrode

The determination of dissolved oxygen in a variety of aqueous environments, such as sea water, blood, sewage, effluents from chemical plants, and soils is of tremendous importance to industry, biomedical and environmental research, and clinical medicine. One of the most common and convenient methods for making such measurements is with the Clark oxygen sensor, which was patented by L.C. Clark, Jr. in 1956.

Differential Pulse Voltammetry

Differential pulse voltammetry (DPV) (also differential pulse polarography, DPP) is a voltammetry method used to make electrochemical measurements and a derivative of linear sweep voltammetry or staircase voltammetry, with a series of regular voltage pulses superimposed on the potential linear sweep or stairsteps. The current is measured immediately before each potential change, and the current difference is plotted as a function of potential. By sampling the current just before the potential is changed, the effect of the charging current can be decreased.

By contrast, in normal pulse voltammetry the current resulting from a series of ever larger potential pulses is compared with the current at a constant 'baseline' voltage. Another type of pulse voltammetry is squarewave voltammetry, which can be considered a special type of differential pulse voltammetry in which equal time is spent at the potential of the ramped baseline and potential of the superimposed pulse.

Electrochemical Cell

The system of this measurement is usually the same as that of standard voltammetry. The potential between the working electrode and the reference electrode is changed as a pulse from an initial potential to an interlevel potential and remains at the interlevel potential for about 5 to 100 milliseconds; then it changes to the final potential, which is different from the initial potential. The pulse is repeated, changing the final potential, and a constant difference is kept between the initial and the interlevel potential. The value of the current between the working electrode and auxiliary electrode before and after the pulse are sampled and their differences are plotted versus potential.

Uses

These measurements can be used to study the redox properties of extremely small amounts of chemicals because of the following two features: 1) in these measurements, the effect of the charging current can be minimized, so high sensitivity is achieved and 2) only faradaic current is extracted, so electrode reactions can be analyzed more precisely.

Characteristics

Differential pulse voltammetry has these characteristics: 1) reversible reactions have symmetric peaks, and irreversible reactions have asymmetric peaks, 2) the peak potential is equal to $E_{1/2}{}^{r}-\Delta E$ in reversible reactions, and the peak current is proportional to the concentration, 3) The detection limit is about 10^{-8} M.

Square Wave Voltammetry

Square wave voltammetry can be used to perform an experiment much faster than normal and differential pulse techniques, which typically run at scan rates of 1 to 10 mV/sec. Square wave voltammetry employs scan rates up to 1 V/sec or faster, allowing much faster determinations. A typical experiment requiring three minutes by normal or differential pulse techniques can be performed in a matter of seconds by square wave voltammetry.

Theory

The waveform used for square wave voltammetry is shown in figure. A symmetrical square wave is superimposed on a staircase waveform where the forward pulse of the square wave (pulse direction same as the scan direction) is coincident with the staircase step. The reverse pulse of the square wave occurs half way through the staircase step.

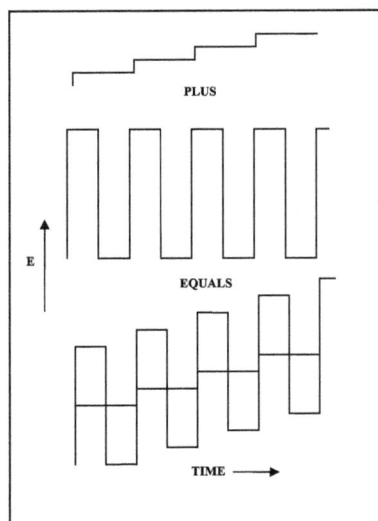

Applied excitation in square wave voltammetry.

The timing and applied potential parameters for square wave voltammetry are depicted in figure. τ is the time for one square wave cycle or one staircase step in seconds. The square wave frequency in Hz is $1/\tau$. E_{sw} is the height of the square wave pulse in mV, where 2 E_sw is equal to the peak-to-peak amplitude. Estep is the staircase step size in mV. The scan rate for a square wave voltammetry experiment can be calculated from the equation:

$$\text{Scan Rate (mV/ sec)} = \frac{E_{step}\ (mV)}{\tau\ (sec)}$$

For example, if E_{step} is 2 mV and τ is 0.01 sec (corresponding to a frequency of 100 Hz), the scan rate would be 200 mV/sec.

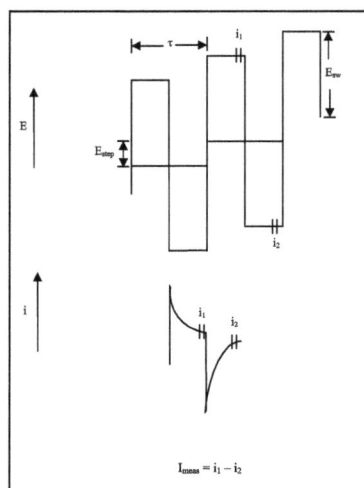

SWV timing relationships.

This scan rate is considerably faster than the 1 to 10 mV/sec rate of other pulse techniques.

The current is sampled twice during each square wave cycle, one at the end of the forward pulse, and again at the end of the reverse pulse. The technique discriminates against charging current by delaying the current measurement to the end of the pulse. The difference current between the two measurements is plotted vs. the potential staircase. Square wave voltammetry yields peaks for faradaic processes, where the peak height is directly proportional to the concentration of the species in solution.

Due to the rapid scan rates possible with square wave voltammetry, the entire voltammogram is recorded on a single mercury drop. Early square wave experiments were limited by the Dropping Mercury Electrode (DME). With a DME, the mercury flows constantly during the experiment, resulting in an electrode of constantly changing surface area. The changing surface area causes slope in the baseline and adds complexity to the experiment theory.

Frequencies used for square wave voltammetry typically range from approximately 1 Hz to 120 Hz. This frequency range allows square wave experiments to be up to 100 times faster than other pulse techniques. The speed of the technique can result in increased sample throughput over previous voltammetric techniques.

Even greater sensitivity can be attained by doing Stripping Square Wave Voltammetry, in which the species of interest is concentrated into the working electrode by electrochemical means before doing the analysis. With a sufficiently long concentration step (deposition), the concentration of the substance will be much higher in the electrode than in the sample solution. If the electrode potential is then scanned, the substance will be stripped form the electrode, causing an increase in cell current as this process occurs. In Anodic Stripping Voltammetry, the concentration is done at a negative potential and the subsequent scan is in the positive direction. During the concentration process, individual metallic species are reduced and deposited into the mercury drop. The subsequent scan direction is positive, causing the individual species to be oxidized back into the solution.

Cathodic stripping is the "mirror image" of anodic stripping analysis. Generally, an insoluble salt of the analyte is concentrated onto the electrode by oxidation. A cathodic (negative going) scan is then applied in which the salt is reduced. Cathodic stripping analysis is best suited to the analysis of those materials that will oxidize into a mercury electrode. Specific examples include selenium, tellurium and the halogens (iodides, bromides, chlorides).

Stripping Voltammetry

Stripping voltammetry is a very sensitive method for the analysis of trace concentrations of electroactive species in solution. Detection limits for metal ions at sub-ppb concentrations have been reported.

There are 3 important parts in a stripping experiment:

- Deposition.

- Quiet time.

- Stripping.

These components can best be explained by discussing the stripping experiment for detection of lead. In this experiment, a mercury working electrode is used - either the Hanging Mercury Drop Electrode (HMDE) or the Thin Mercury Film Electrode (TMFE). The TMFE is made by a depositing a mercury film on the surface of a glassy carbon electrode, typically during the deposition step.

Deposition E (mV)	-800	Deposition Time (Sec)	30
Use Initial E as Deposition E	☐	Quiet Time (Sec)	2
Stir during Deposition	☑	Purge during Deposition	☐

Parameters for deposition step when using the CGME.

Deposition E (mV)	-800	Deposition Time (Sec)	30
Use Initial E as Deposition E	☐	Quiet Time (Sec)	2
Rotate during Deposition	☑	Purge during Deposition	☐

Parameters for deposition stepwhen using the RDE-2.

The parameters for the deposition step are shown in the earlier figures. During the deposition step, the potential applied to the mercury electrode is held at a value (Deposition E) at which the lead ions are reduced to lead metal for a pre-determined time period (Deposition Time). If the Use Initial E as Deposition E is checked, the Initial E of the stripping step is used as the Deposition Potential. The metallic lead then amalgamates with the mercury electrode (when the TMFE is used, mercuric ions are generally added to the solution, and mercury metal is codeposited with the lead during the deposition step). The effect of this amalgamation is to concentrate the lead in the mercury electrode, and hence the concentration of lead in the electrode is much greater (typically 2 or 3 orders of magnitude) than the concentration of lead in the solution (consequently, the deposition step is often called the preconcentration or accumulation step). The efficiency of the deposition can be increased by stirring either the solution (when using the CGME) or rotating the electrode (when using the RDE-2). Stirring (for the CGME), rotating (for the RDE-2), and purging during this step can be controlled remotely from the software by checking the Stir/Rotate during Deposition and Purge during Deposition boxes, respectively. Cell Stand in Setup/Manual Settings (I/O) in the Experiment menu must be set to CGME SMDE Mode when using the CGME and to RDE-2 when using the RDE-2.

After the deposition step, the stirring is stopped, and the system is allowed to reach equilibrium. This is the Quiet Time, which is typically 10 - 15 s.

During the stripping step, the applied potential is scanned in a positive direction, and the lead in the mercury electrode is oxidized back to lead ions in solution; that is, the lead is "stripped" from the electrode. The potential at which the stripping occurs is related to the redox potential of the analyte, and hence the potential of the current peak on the stripping step can be used to identify the analyte. The magnitude of the current of the stripping peak is proportional to the concentration of the analyte in the mercury electrode. Since the concentration of the analyte in the electrode is related to its concentration in solution, the stripping peak current is therefore proportional to the solution concentration.

A number of different wave forms have been used for the stripping step, including linear sweep voltammetry (LSSV), differential pulse voltammetry (DPSV), and square wave voltammetry (SWSV). SWSV and DPSV are more commonly used, due to their lower detection limits.

As noted above, it is the concentration of lead in the mercury electrode that is directly measured in the stripping step rather than the concentration of lead in solution. The electrode concentration can be increased by increasing the Deposition Time and the rate of stirring. The values required for these two parameters depends on the sensitivity of the mercury electrode, which is determined by the surface area to volume ratio (i.e., how many of the deposited lead atoms are on the mercury surface and hence are detectable in the stripping step). This ratio is considerably higher for the TMFE, so a shorter Deposition Time is required. In addition, faster stirring can be used with the TMFE due to the relative mechanical instability of the HMDE (i.e., the mercury drop can fall off if the stirring is too fast). The signal resolution is also better with the TMFE, which can be important if there is more than one metal ion present.

However, the greater sensitivity of the TMFE can also be a disadvantage, since the solubility limit of the metal in the mercury can be exceeded more readily. This can lead to the formation of intermetallic compounds, which can affect the accuracy of the experimental results (due to e.g., shifts in the stripping potentials and depression of the stripping currents). One pair of metals that readily combine is zinc and copper.

In order to be of use as a quantitative analytical technique, the results of a stripping experiments must be reproducible. Therefore, the experimental conditions must be reproducible. A second disadvantage of the TMFE is the relatively poor reproducibility of the film. Since the film is deposited on the surface of a glassy carbon electrode, it is sensitive to the microstructure of the glassy carbon surface, which can be affected by the method used to prepare the surface. In contrast, an HMDE is highly reproducible. Whatever the chosen mercury electrode, great care must be taken in sample preparation, cleaning of glassware, etc. The rate of stirring during the deposition step must also remain constant.

The above method is called anodic stripping voltammetry (ASV), since the stripping current is anodic. This method can be used for metal ions that can be readily reduced to the metallic state and reoxidized - about 20 metal ions, including lead, copper, cadmium, and zinc. This is not as many as can be detected using atomic absorption spectroscopy (AAS), although the sensitivity of ASV is comparable with, and sometimes better than AAS. The advantage of ASV over AAS is its ability to detect several metal ions simultaneously. In addition, different oxidation states of a given metal can be detected (e.g., arsenic and antimony).

Other stripping voltammetric techniques include cathodic stripping voltammetry (CSV) and adsorptive stripping voltammetry (AdSV). The basis for CSV is the oxidation of mercury followed by the formation of an insoluble film of HgL (L is the analyte) on the surface of the mercury electrode during the deposition step. CSV is most commonly used for detection of sulfur-containing molecules (e.g., thiols, thioureas, and thioamides), but it has also been used for molecules such as riboflavin and nucleic acid bases (e.g., adenine and cytosine).

AdSV is different from ASV and CSV in that the deposition step is non-electrolytic, and occurs via the adsorption of molecules on the surface of the working electrode (the HMDE is most commonly used). The stripping step can be either anodic or cathodic. AdSV has been used for organic molecules (e.g., dopamine, chlorpromazine, erythromycin, dibutone, and ametryne) and for metal complexes of metals not amenable to detection by ASV (e.g., cobalt and nickel).

Hydrodynamic Voltammetry

Hydrodynamic voltammetry is a form of voltammetry in which the analyte solution flows relative to a working electrode. In many voltammetry techniques, the solution is intentionally left still to allow diffusion controlled mass transfer. When a solution is made to flow, through stirring or some other physical mechanism, it is very important to the technique to achieve a very controlled flux or mass transfer in order to obtain predictable results. These methods are types of electrochemical studies which use potentiostats to investigate reaction mechanisms related to redox chemistry among other chemical phenomenon.

Structure

Most experiment involve a three electrode setup but the configuration of the setup varies widely. All cell configurations create a laminar flow of solution across the working electrode(s) producing a steady-state current determined by solution flow rather than diffusion. The current resulting can be mathematically predicted and modeled. Among the most common hydrodynamic setup involves the working electrodes rotating to create a laminar flow of solution across the electrode surface. Both rotating disk electrodes (RDE) and rotating ring-disk electrodes (RRDE) are examples where the working electrode rotates. Other configurations, such as flow cells, use pumps to direct solution at or across the working electrode(s).

Distinction

Hydrodynamic techniques are distinct from still and unstirred experiments such as cyclic voltammetry where the stead-state current is limited by the diffusion of substrate. Experiments are not however limited to linear sweep voltammetry. The configuration of many cells takes the substrate from one working electrode across another, RRDE for example. The potential of one electrode can be varied as the other is held constant or varied. The flow rate can also be varied to adjust the temporal gap the substrates experiences between working electrodes.

Linear Sweep Voltammetry

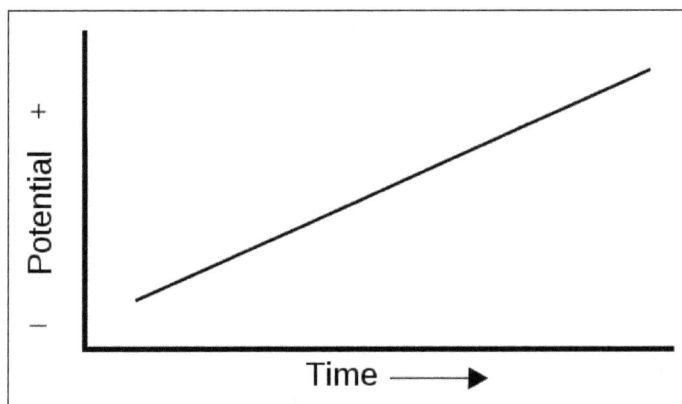

Linear potential sweep.

Linear sweep voltammetry is a voltammetric method where the current at a working electrode is measured while the potential between the working electrode and a reference electrode is swept linearly in time. Oxidation or reduction of species is registered as a peak or trough in the current signal at the potential at which the species begins to be oxidized or reduced.

Experimental Method

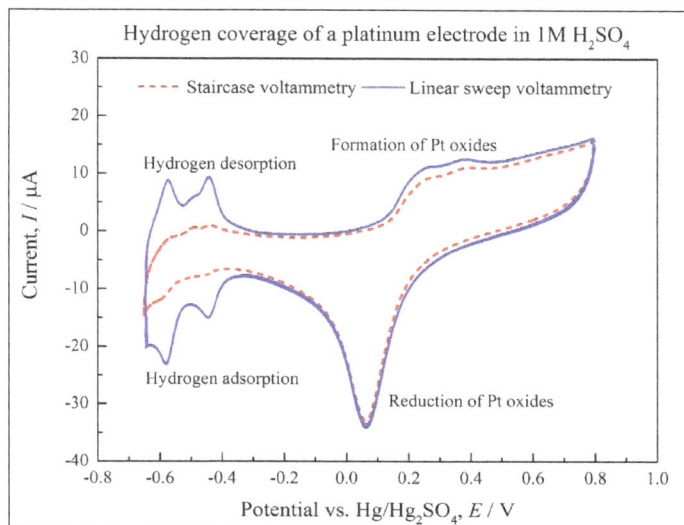

Comparison of the current response of a platinum disc electrode in 1 M sulphuric acid given by linear sweep cyclic voltammetry and staircase cyclic voltammetry methods. Staircase voltammetry suppresses the non-faradaic adsorption of hydrogen.

The experimental setup for linear sweep voltammetry utilizes a potentiostat and a three-electrode setup to deliver a potential to a solution and monitor its change in current. The three-electrode setup consists of a working electrode, an auxiliary electrode, and a reference electrode. The potentiostat delivers the potentials through the three-electrode setup. A potential, E, is delivered through the working electrode. The slope of the potential vs. time graph is called the scan rate and can range from mV/s to 1,000,000 V/s. The working electrode is one of the electrodes at which the oxidation/reduction reactions occur—the processes that occur at this electrode are the ones being monitored. The auxiliary electrode (or counter electrode) is the one at which a process opposite from the one taking place at the working electrode occurs. The processes at this electrode are not monitored. The equation below gives an example of a reduction occurring at the surface of the working electrode. E_s is the reduction potential of A (if the electrolyte and the electrode are in their standard conditions, then this potential is a standard reduction potential). As E approaches E_s the current on the surface increases and when E=E_s then the concentration of [A] = [A⁻] at the surface. As the molecules on the surface of the working electrode are oxidized/reduced they move away from the surface and new molecules come into contact with the surface of the working electrode. The flow of electrons into or out of the electrode causes the current. The current is a direct measure of the rate at which electrons are being exchanged through the electrode-electrolyte interface. When this rate becomes higher than the rate at which the oxidizing or reducing species can diffuse from the bulk of the electrolyte to the surface of the electrode, the current reaches a plateau or exhibits a peak.

$$A + e^- \rightleftharpoons A^-, E_s = 0.00V$$

Reduction of molecule A at the surface of the working electrode.

The auxiliary and reference electrode work in unison to balance out the charge added or removed by the working electrode. The auxiliary electrode balances the working electrode, but in order to know how much potential it has to add or remove it relies on the reference electrode. The reference electrode has a known reduction potential. The auxiliary electrode tries to keep the reference electrode at a certain reduction potential and to do this it has to balance the working electrode.

Characterization

Linear sweep voltammetry can identify unknown species and determine the concentration of solutions. E1/2 can be used to identify the unknown species while the height of the limiting current can determine the concentration. The sensitivity of current changes vs. voltage can be increased by increasing the scan rate. Higher potentials per second result in more oxidation/reduction of a species at the surface of the working electrode.

Variations

For reversible reactions cyclic voltammetry can be used to find information about the forward reaction and the reverse reaction. Like linear sweep voltammetry, cyclic voltammetry applies a linear potential over time and at a certain potential the potentiostat will reverse the potential applied and sweep back to the beginning point. Cyclic voltammetry provides information about the oxidation and reduction reactions.

Applications

While cyclic voltammetry is applicable to most cases where linear sweep voltammetry is used, there are some instances where linear sweep voltammetry is more useful. In cases where the reaction is irreversible cyclic voltammetry will not give any additional data that linear sweep voltammetry would give us. In one example, linear voltammetry was used to examine direct methane production via a biocathode. Since the production of methane from CO_2 is an irreversible reaction, cyclic voltammetry did not present any distinct advantage over linear sweep voltammetry. This group found that the biocathode produced higher current densities than a plain carbon cathode and that methane can be produced from a direct electric current without the need of hydrogen gas.

Cyclic Voltammetry

Cyclic voltammetry (CV) is a type of potentiodynamic electrochemical measurement. In a cyclic voltammetry experiment, the working electrode potential is ramped linearly versus time. Unlike in linear sweep voltammetry, after the set potential is reached in a CV experiment, the working electrode's potential is ramped in the opposite direction to return to the initial potential. These cycles of ramps in potential may be repeated as many times as needed. The current at the working electrode is plotted versus the applied voltage (that is, the working electrode's potential) to give the cyclic voltammogram trace. Cyclic voltammetry is generally used to study the electrochemical properties of an analyte in solution or of a molecule that is adsorbed onto the electrode.

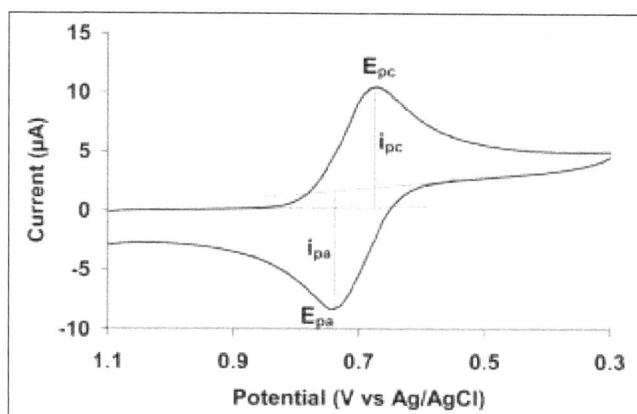

Typical cyclic voltammogram where i_{pc} and i_{pa} show the peak cathodic and anodic current respectively for a reversible reaction.

Experimental Method

In cyclic voltammetry, the electrode potential ramps linearly versus time in cyclical phases. The rate of voltage change over time during each of these phases is known as the experiment's scan rate (V/s). The potential is measured between the working electrode and the reference electrode, while the current is measured between the working electrode and the counter electrode. These data are plotted as current (i) versus applied potential (E, often referred to as just 'potential'). In the figure above, during the initial forward scan (from t_0 to t_1) an increasingly reducing potential is applied; thus the cathodic current will, at least initially, increase over this time period assuming that there are reducible analytes in the system. At some point after the reduction potential of the analyte is reached, the cathodic current will decrease as the concentration of reducible analyte is depleted. If the redox couple is reversible then during the reverse scan (from t_1 to t_2) the reduced analyte will start to be re-oxidized, giving rise to a current of reverse polarity (anodic current) to before. The more reversible the redox couple is, the more similar the oxidation peak will be in shape to the reduction peak. Hence, CV data can provide information about redox potentials and electrochemical reaction rates.

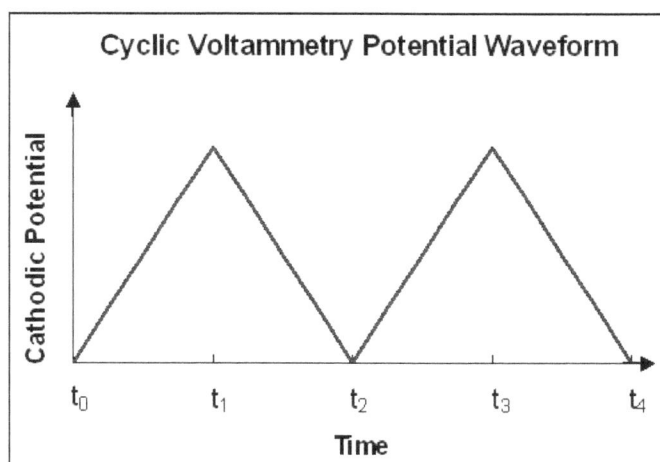

Cyclic voltammetry waveform.

For instance, if the electron transfer at the working electrode surface is fast and the current is limited by the diffusion of analyte species to the electrode surface, then the peak current will be proportional to the square root of the scan rate. This relationship is described by the Randles–Sevcik equation. In this situation, the CV experiment only samples a small portion of the solution, i.e., the diffusion layer at the electrode surface.

Characterization

The utility of cyclic voltammetry is highly dependent on the analyte being studied. The analyte has to be redox active within the potential window to be scanned.

The Analyte is in Solution

Reversible Couples

Often the analyte displays a reversible CV wave, which is observed when all of the initial analyte can be recovered after a forward and reverse scan cycle. Although such reversible couples are simpler to analyze, they contain less information than more complex waveforms.

The waveform of even reversible couples is complex owing to the combined effects of polarization and diffusion. The difference between the two peak potentials (E_p), ΔE_p, is of particular interest.

$$\Delta E_p = E_{pa} - E_{pc} > 0$$

This difference mainly results from the effects of analyte diffusion rates. In the ideal case of a reversible 1e- couple, ΔE_p is 57 mV and the full-width half-max of the forward scan peak is 59 mV. Typical values observed experimentally are greater, often approaching 70 or 80 mV. The waveform is also affected by the rate of electron transfer, usually discussed as the activation barrier for electron transfer. A theoretical description of polarization overpotential is in part described by the Butler–Volmer equation and Cottrell equation equations. In an ideal system the relationships

reduces to $E_{pa} - E_{pc} = \dfrac{56.5 \text{ mV}}{n}$ for an n electron process.

Focusing on current, reversible couples are characterized by $i_{pa} / i_{pc} = 1$.

When a reversible peak is observed, thermodynamic information in the form of a half cell potential $E^\circ_{1/2}$ can be determined. When waves are semi-reversible (i_{pa}/i_{pc} is close but not equal to 1), it may be possible to determine even more specific information.

Nonreversible Couples

Many redox processes observed by CV are quasi-reversible or non-reversible. In such cases the thermodynamic potential $E^\circ_{1/2}$ is often deduced by simulation. The irreversibility is indicated by $i_{pa} / i_{pc} \neq 1$. Deviations from unity are attributable to a subsequent chemical reaction that is triggered by the electron transfer. Such EC processes can be complex, involving isomerization, dissociation, association, etc.

The Analyte is Adsorbed onto the Electrode Surface

Adsorbed species give simple voltammetric responses: ideally, at slow scan rates, there is no peak separation, the peak width is 90mV for a one-electron redox couple, and the peak current and peak area are proportional to scan rate (observing that the peak current is proportional to scan rate proves that the redox species that gives the peak is actually immobilised). The effect of increasing the scan rate can be used to measure the rate of interfacial electron transfer and the rates of reactions that are coupled to electron transfer. This technique has been useful to study redox proteins, some of which readily adsorb on various electrode materials, but the theory for biological and non-biological redox molecules is the same.

Experimental Setup

CV experiments are conducted on a solution in a cell fitted with electrodes. The solution consists of the solvent, in which is dissolved electrolyte and the species to be studied.

The Cell

A standard CV experiment employs a cell fitted with three electrodes: reference electrode, working electrode, and counter electrode. This combination is sometimes referred to as a three-electrode setup. Electrolyte is usually added to the sample solution to ensure sufficient conductivity. The solvent, electrolyte, and material composition of the working electrode will determine the potential range that can be accessed during the experiment.

The electrodes are immobile and sit in unstirred solutions during cyclic voltammetry. This "still" solution method gives rise to cyclic voltammetry's characteristic diffusion-controlled peaks. This method also allows a portion of the analyte to remain after reduction or oxidation so that it may display further redox activity. Stirring the solution between cyclic voltammetry traces is important in order to supply the electrode surface with fresh analyte for each new experiment. The solubility of an analyte can change drastically with its overall charge; as such it is common for reduced or oxidized analyte species to precipitate out onto the electrode. This layering of analyte can insulate the electrode surface, display its own redox activity in subsequent scans, or otherwise alter the electrode surface in a way that affects the CV measurements. For this reason it is often necessary to clean the electrodes between scans.

Common materials for the working electrode include glassy carbon, platinum, and gold. These electrodes are generally encased in a rod of inert insulator with a disk exposed at one end. A regular working electrode has a radius within an order of magnitude of 1 mm. Having a controlled surface area with a well-defined shape is necessary for being able to interpret cyclic voltammetry results.

To run cyclic voltammetry experiments at very high scan rates a regular working electrode is insufficient. High scan rates create peaks with large currents and increased resistances, which result in distortions. Ultramicroelectrodes can be used to minimize the current and resistance.

The counter electrode, also known as the auxiliary or second electrode, can be any material that conducts current easily and will not react with the bulk solution. Reactions occurring at the counter electrode surface are unimportant as long as it continues to conduct current well. To maintain the observed current the counter electrode will often oxidize or reduce the solvent or bulk electrolyte.

Solvents

CV can be conducted using a variety of solutions. Using typical electrodes, solvents dissolve not only the analyte, often at mM levels, but also electrolyte, generally at much higher concentrations. For aqueous solutions, these requirements are trivial, but for nonaqueous solutions, the choices of suitable solvents are fewer.

Electrolyte

The electrolyte ensures good electrical conductivity and minimizes iR drop such that the recorded potentials correspond to actual potentials. For aqueous solutions, many electrolytes are available, but typical ones are alkali metal salts of perchlorate and nitrate. In nonaqueous solvents, the range of electrolytes is more limited, and a popular choice is tetrabutylammonium hexafluorophosphate.

Related Potentiometric Techniques

Potentiodynamic techniques also exist that add low-amplitude AC perturbations to a potential ramp and measure variable response in a single frequency (AC voltammetry) or in many frequencies simultaneously (potentiodynamic electrochemical impedance spectroscopy). The response in alternating current is two-dimensional, characterized by both amplitude and phase. These data can be analyzed to determine information about different chemical processes (charge transfer, diffusion, double layer charging, etc.). Frequency response analysis enables simultaneous monitoring of the various processes that contribute to the potentiodynamic AC response of an electrochemical system.

Whereas cyclic voltammetry is not a hydrodynamic technique, useful electrochemical methods are. In such cases, flow is achieved at the electrode surface by stirring the solution, pumping the solution, or rotating the electrode as is the case with rotating disk electrodes and rotating ring-disk electrodes. Such techniques target steady state conditions and produce waveforms that appear the same when scanned in either the positive or negative directions, thus limiting them to linear sweep voltammetry.

Applications

Cyclic voltammetry (CV) has become an important and widely used electroanalytical technique in many areas of chemistry. It is often used to study a variety of redox processes, to determine the stability of reaction products, the presence of intermediates in redox reactions, electron transfer kinetics, and the reversibility of a reaction. CV can also be used to determine the electron stoichiometry of a system, the diffusion coefficient of an analyte, and the formal reduction potential of an analyte, which can be used as an identification tool. In addition, because concentration is proportional to current in a reversible, Nernstian system, the concentration of an unknown solution can be determined by generating a calibration curve of current vs. concentration.

In cellular biology it is used to measure the concentrations in living organisms. In organometallic chemistry, it is used to evaluate redox mechanisms.

Measuring Antioxidant Capacity

Cyclical voltammetry can be used to determine the antioxidant capacity in food and even skin. Low molecular weight antioxidants, molecules that prevent other molecules from being oxidized by acting as reducing agents, are important in living cells because they inhibit cell damage or death caused by oxidation reactions that produce radicals. Examples of antioxidants include flavonoids, whose antioxidant activity is greatly increased with more hydroxyl groups. Because traditional methods to determine antioxidant capacity involve tedious steps, techniques to increase the rate of the experiment are continually being researched. One such technique involves cyclic voltammetry because it can measure the antioxidant capacity by quickly measuring the redox behavior over a complex system without the need to measure each component's antioxidant capacity. Furthermore, antioxidants are quickly oxidized at inert electrodes, so the half-wave potential can be utilized to determine antioxidant capacity. It is important to note that whenever cyclic voltammetry is utilized, it is usually compared to spectrophotometry or High Performance Liquid Chromotography (HPLC). Applications of the technique extend to food chemistry, where it is used to determine the antioxidant activity of red wine, chocolate, and hops. Additionally, it even has uses in the world of medicine in that it can determine antioxidants in the skin.

Evaluation of a Technique

The technique being evaluated uses voltammetric sensors combined in an electronic tongue to observe the antioxidant capacity in red wines. These Electronic Tongues (ETs) consist of multiple sensing units like voltammetric sensors, which will have unique responses to certain compounds. This approach is optimal to use since samples of high complexity can be analyzed with high cross-selectivity. Thus, the sensors can be sensitive to pH and antioxidants. As usual, the voltage in the cell was monitored using a working electrode and a reference electrode (Silver/Silver Chloride electrode). Furthermore, a platinum counter electrode allows the current to continue to flow during the experiment. The Carbon Paste Electrodes sensor (CPE) and the Graphite-Epoxy Composite (GEC) electrode are tested in a saline solution before the scanning of the wine so that a reference signal can be obtained. The wines are then ready to be scanned, once with CPE and once with GEC. While cyclic voltammetry was successfully used to generate currents using the wine samples, the signals were complex and needed an additional extraction stage. It was found that the ET method could successfully analyze wine's antioxidant capacity as it agreed with traditional methods like TEAC, Folin-Ciocalteu, and I280 indexes. Additionally, the time was reduced, the sample did not have to be pretreated, and other reagents were unnecessary, all of which diminshed the popularity of traditional methods. Thus, cyclic voltammetry successfully determines the antioxidant capacity and even improves previous results.

Antioxidant Capacity of Chocolate and Hops

The phenolic antioxidants for cocoa powder, dark chocolate, and milk chocolate can also be determined via cyclic voltammetry. In order to achieve this, the anodic peaks are calculated and analyzed with the knowledge that the first and third anodic peaks can be assigned to the first and second oxidation of flavonoids, while the second anodic peak represents phenolic acids. Using the graph produced by cyclic voltammetry, the total phenolic and flavonoid content can be deduced in each of the three samples. It was observed that cocoa powder and dark chocolate had the highest antioxidant capacity since they had high total phenolic and flavonoid content. Milk chocolate had the lowest capacity as it had the lowest phenolic and flavonoid content. While the antioxidant content was given using the cyclic voltammetry anodic peaks, HPLC must then be used to determine the purity of catechins and procyanidin in cocoa powder, dark chocolate, and milk chocolate.

Hops, the flowers used in making beer, contain antioxidant properties due to the presence of flavonoids and other polyphenolic compounds. In this cyclic voltammetry experiment, the working electrode voltage was determined using a ferricinium/ferrocene reference electrode. By comparing different hop extract samples, it was observed that the sample containing polyphenols that were oxidized at less positive potentials proved to have better antioxidant capacity.

Polarography

Polarography is also called as polarographic analysis, or voltammetry, in analytic chemistry, an electrochemical method of analyzing solutions of reducible or oxidizable substances. It was invented by a Czech chemist, Jaroslav Heyrovský, in 1922.

In general, polarography is a technique in which the electric potential (or voltage) is varied in a regular manner between two sets of electrodes (indicator and reference) while the current is monitored. The shape of a polarogram depends on the method of analysis selected, the type of indicator electrode used, and the potential ramp that is applied. The figure shows five selected methods of polarography; the potential ramps are applied to a mercury indicator electrode, and the shapes of the resulting polarograms are compared.

The majority of the chemical elements can be identified by polarographic analysis, and the method is applicable to the analysis of alloys and to various inorganic compounds. Polarography is also used to identify numerous types of organic compounds and to study chemical equilibria and rates of reactions in solutions.

The solution to be analyzed is placed in a glass cell containing two electrodes. One electrode consists of a glass capillary tube from which mercury slowly flows in drops, and the other is commonly a pool of mercury. The cell is connected in series with a galvanometer (for measuring the flow of current) in an electrical circuit that contains a battery or other source of direct current and a device for varying the voltage applied to the electrodes from zero up to about two volts. With the dropping mercury electrode connected (usually) to the negative side of the polarizing voltage, the voltage is increased by small increments, and the corresponding current is observed on the galvanometer. The current is very small until the applied voltage is increased to a value large enough to cause the substance being determined to be reduced at the dropping mercury electrode. The current increases rapidly at first as the applied voltage is increased above this critical value but gradually attains a limiting value and remains more or less constant as the voltage is increased further. The critical voltage required to cause the rapid increase in current is characteristic of, and also serves to identify, the substance that is being reduced (qualitative analysis). Under proper conditions the constant limiting current is governed by the rates of diffusion of the reducible substance up to the surface of the mercury drops, and its magnitude constitutes a measure of the concentration of the reducible substance (quantitative analysis). Limiting currents also result from the oxidation of certain oxidizable substances when the dropping electrode is the anode.

When the solution contains several substances that are reduced or oxidized at different voltages, the current-voltage curve shows a separate current increase (polarographic wave) and limiting current for each. The method is thus useful in detecting and determining several substances simultaneously and is applicable to relatively small concentrations—e.g., 10−6 up to about 0.01 mole per litre, or approximately 1 to 1,000 parts per 1,000,000.

AMPEROMETRIC TITRATION

Amperometric titration refers to a class of titrations in which the equivalence point is determined through measurement of the electric current produced by the titration reaction. It is a form of quantitative analysis.

A solution containing the analyte, A, in the presence of some conductive buffer. If an electrolytic potential is applied to the solution through a working electrode, then the measured current depends (in part) on the concentration of the analyte. Measurement of this current can be used to determine the concentration of the analyte directly; this is a form of amperometry. However, the difficulty is that the measured current depends on several other variables, and it is not always possible to control all of them adequately. This limits the precision of direct amperometry.

If the potential applied to the working electrode is sufficient to reduce the analyte, then the concentration of analyte close to the working electrode will decrease. More of the analyte will slowly diffuse into the volume of solution close to the working electrode, restoring the concentration. If the potential applied to the working electrode is great enough (an overpotential), then the concentration of analyte next to the working electrode will depend entirely on the rate of diffusion. In such a case, the current is said to be diffusion limited. As the analyte is reduced at the working electrode, the concentration of the analyte in the whole solution will very slowly decrease; this depends on the size of the working electrode compared to the volume of the solution.

What happens if some other species which reacts with the analyte (the titrant) is added? (For instance, chromate ions can be added to oxidize lead ions.) After a small quantity of the titrant (chromate) is added, the concentration of the analyte (lead) has decreased due to the reaction with chromate. The current from the reduction of lead ion at the working electrode will decrease. The addition is repeated, and the current decreases again. A plot of the current against volume of added titrant will be a straight line.

After enough titrant has been added to react completely with the analyte, the excess titrant may itself be reduced at the working electrode. Since this is a different species with different diffusion characteristics (and different half-reaction), the slope of current versus added titrant will have a different slope after the equivalence point. This change in slope marks the equivalence point, in the same way that, for instance, the sudden change in pH marks the equivalence point in an acid-base titration.

The electrode potential may also be chosen such that the titrant is reduced, but the analyte is not. In this case, the presence of excess titrant is easily detected by the increase in current above background (charging) current.

Advantages

The chief advantage over direct amperometry is that the magnitude of the measured current is of interest only as an indicator. Thus, factors that are of critical importance to quantitative amperometry, such as the surface area of the working electrode, completely disappear from amperometric titrations.

The chief advantage over other types of titration is the selectivity offered by the electrode potential, as well as by the choice of titrant. For instance, lead ion is reduced at a potential of -0.60 V (relative to the saturated calomel electrode), while zinc ions are not; this allows the determination of lead in the presence of zinc. Clearly this advantage depends entirely on the other species present in the sample.

ELECTROCHEMICAL STRIPPING ANALYSIS

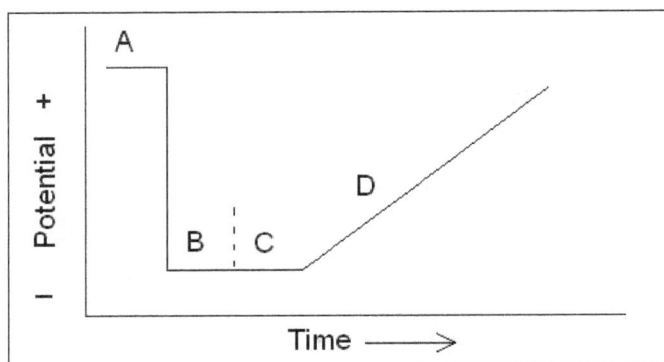

A: cleaning step, B: electroplating step, C: equilibration step, D: stripping step.

Electrochemical stripping analysis is a set of analytical chemistry methods based on voltammetry or potentiometry that are used for quantitative determination of ions in solution. Stripping voltammetry (anodic, cathodic and adsorptive) have been employed for analysis of organic molecules as well as metal ions. Carbon paste, glassy carbon paste, and glassy carbon electrodes when modified are termed as chemically modified electrodes and have been employed for the analysis of organic and inorganic compounds.

Stripping analysis is an analytical technique that involves (i) preconcentration of a metal phase onto a solid electrode surface or into Hg (liquid) at negative potentials and (ii) selective oxidation of each metal phase species during an anodic potential sweep. Stripping analysis has the following properties: sensitive and reproducible (RSD<5%) method for trace metal ion analysis in aqueous media, 2) concentration limits of detection for many metals are in the low ppb to high ppt range (S/N=3) and this compares favorably with AAS or ICP analysis, field deployable instrumentation that is inexpensive, approximately 12-15 metal ions can be analyzed for by this method. The stripping peak currents and peak widths are a function of the size, coverage and distribution of the metal phase on the electrode surface (Hg or alternate).

Anodic Stripping Voltammetry

Anodic stripping voltammetry is a voltammetric method for quantitative determination of specific ionic species. The analyte of interest is electroplated on the working electrode during a deposition step, and oxidized from the electrode during the stripping step. The current is measured during the stripping step. The oxidation of species is registered as a peak in the current signal at the potential at which the species begins to be oxidized. The stripping step can be either linear, staircase, squarewave, or pulse.

Anodic stripping voltammetry usually incorporates three electrodes, a working electrode, auxiliary electrode (sometimes called the counter electrode), and reference electrode. The solution being analyzed usually has an electrolyte added to it. For most standard tests, the working electrode is a bismuth or mercury film electrode (in a disk or planar strip configuration). The mercury film forms an amalgam with the analyte of interest, which upon oxidation results in a sharp peak, improving resolution between analytes. The mercury film is formed over a glassy carbon electrode. A mercury drop electrode has also been used for much the same reasons. In cases where the analyte of interest has an oxidizing potential above that of mercury, or where a mercury electrode would be otherwise unsuitable, a solid, inert metal such as silver, gold, or platinum may also be used.

Anodic stripping voltammetry usually incorporates 4 steps if the working electrode is a mercury film or mercury drop electrode and the solution incorporates stirring. The solution is stirred during the first two steps at a repeatable rate. The first step is a cleaning step; in the cleaning step, the potential is held at a more oxidizing potential than the analyte of interest for a period of time in order to fully remove it from the electrode. In the second step, the potential is held at a lower potential, low enough to reduce the analyte and deposit it on the electrode. After the second step, the stirring is stopped, and the electrode is kept at the lower potential. The purpose of this third step is to allow the deposited material to distribute more evenly in the mercury. If a solid inert electrode is used, this step is unnecessary. The last step involves raising the working electrode to a higher potential (anodic), and stripping (oxidizing) the analyte. As the analyte is oxidized, it gives off electrons which are measured as a current.

Anodic stripping voltammetry can detect µg/L concentrations of analyte. This method has an excellent detection limit (typically 10^{-9} - 10^{-10} M).

Cathodic Stripping Voltammetry

Cathodic stripping voltammetry is a voltammetric method for quantitative determination of specific ionic species. It is similar to the trace analysis method anodic stripping voltammetry, except that for the plating step, the potential is held at an oxidizing potential, and the oxidized species are stripped from the electrode by sweeping the potential negatively. This technique is used for ionic species that form insoluble salts and will deposit on or near the anodic, working electrode during deposition. The stripping step can be either linear, staircase, squarewave, or pulse.

Adsorptive Stripping Voltammetry

Adsorptive stripping voltammetry is similar to anodic and cathodic stripping voltammetry except that the preconcentration step is not controlled by electrolysis. The preconcentration step in adsorptive stripping voltammetry is accomplished by adsorption on the working electrode surface, or by reactions with chemically modified electrodes.

CHRONOAMPEROMETRY

Chronoamperometry is an electrochemical technique in which the potential of the working electrode is stepped and the resulting current from faradaic processes occurring at the electrode (caused by the potential step) is monitored as a function of time. The functional relationship between current response and time is measured after applying single or double potential step to the working electrode of the electrochemical system. Limited information about the identity of the electrolyzed species can be obtained from the ratio of the peak oxidation current versus the peak reduction current. However, as with all pulsed techniques, chronoamperometry generates high charging currents, which decay exponentially with time as any RC circuit. The Faradaic current - which is due to electron transfer events and is most often the current component of interest - decays as described in the Cottrell equation. In most electrochemical cells this decay is much slower than the charging decay-cells with no supporting electrolyte are notable exceptions. Most commonly a three electrode system is used. Since the current is integrated over relatively longer time intervals, chronoamperometry gives a better signal to noise ratio in comparison to other amperometric techniques.

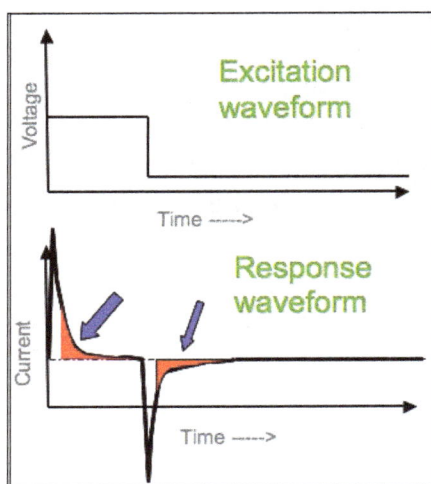

Double-pulsed chronoamperometry waveform showing integrated region for charge determination.

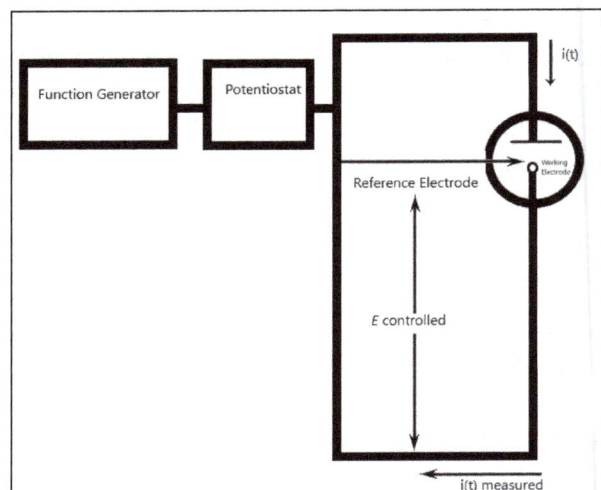

Scheme of Chronoamperometry Instrument.

There are two types of chronoamperometry that are commonly used, controlled-potential chronoamperometry and controlled-current chronoamperometry. Before running controlled-potential chronoamperometry, cyclic voltametries are run to determine the reduction potential of the analytes. Generally, chronoamperometry uses fixed area electrodes, which is suitable for studying electrode processes of coupled chemical reactions, especially the reaction mechanism of organic electrochemistry.

Anthracene in deoxygenated dimethylformamide (DMF) will be reduced (An + e$^-$ -> An$^-$) at the electrode surface that is at a certain negative potential. The reduction will be diffusion-limited, thereby causing the current to drop in time (proportional to the diffusion gradient that is formed by diffusion).

You can do this experiment several times increasing electrode potentials from low to high. (In between the experiments, the solution should be stirred.) When you measure the current i(t) at a certain fixed time point τ after applying the voltage, you will see that at a certain moment the current i(τ) does not rise anymore; you have reached the mass-transfer-limited region. This means that anthracene arrives as fast as diffusion can bring it to the electrode.

Application

Controlled-potential (Bulk) Electrolysis

One of the application of chronoamperometry is controlled-potential (bulk) electrolysis, which is also known as potentiostatic coulometry. During this process, a constant potential is applied to the working electrode and current is monitored over time. The analyte in one oxidation state will be oxidized or reduced to another oxidation state. The current will decrease to the base line (approaching zero) as the analyte is consumed. This process shows the total charge (in coulomb) that flows in the reaction. Total charge (n value) is calculated by integration of area under the current plot and the application of the Faraday's law.

The cell for controlled-potential (bulk) electrolysis is usually a two-compartment (divided) cell, contained a carbon rod auxiliary anode and is separated from the cathode compartment by a coarse glass frit and methyl cellulose solvent electrolyte plug. The reason for the two compartment cell is to separate cathodic and anodic reaction. The working electrode for bulk electrolysis could be a RVC disk, which has larger surface area to increase the rate of the reaction.

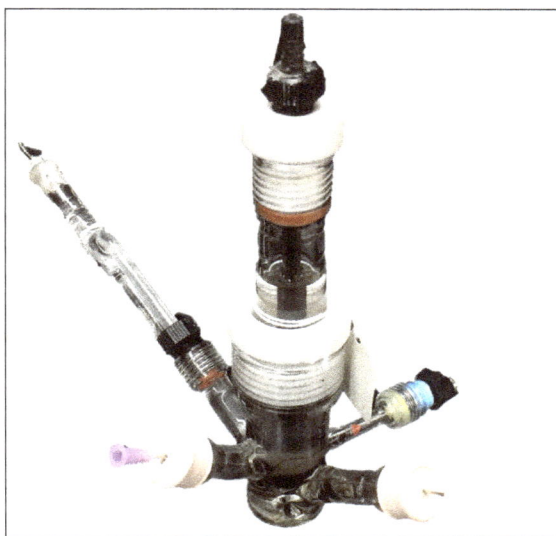

Cell of Controlled-Potential Electrolysis.

Controlled-potential electrolysis is normally utilized with cyclic voltammetry. Cyclic voltammetry is capable to analysis the electrochemical behavior of the analyte or the reaction. For instance, cyclic voltammetry could tell us the cathodic potential of an analyte. Since the cathodic potential of this analyte is obtained, controlled-potential electrolysis could hold this constant potential for the reaction to happen.

Double Potential Step Chronoamperometry (DPSCA)

DPSCA is the technique whose working electrode is applied by the potential stepping forward for a certain period of time and backward for a period of time. The current is monitored and plotted with respect to time. This method starts with an induction period. In this period, several initial conditions will be applied to the electrochemical cell so that cell is able to equilibrate to those conditions. The working electrode potential will be held at the initial potential under these conditions for a specified period (i.e. usually 3 seconds). When the induction period is over, the working cells switch to another potential for a certain amount of time. After the first step is completed, the working electrode's potential will stepped back, usually to the potential prior to the forward step. The whole experiment ends with a relaxation period. Under this period, the default condition involves holding the working electrode potential of initial state for another approximate 1 seconds. When the relaxation period is over, the post experiment idle conditions will be applied to the cell so that the instrument can return to the idle state1. After plotting the current as a function of time, a chronoamperogram will occur and it can also be used to generate Cottrell plots.

Cell of Cyclic Voltammetry.

Other Two Methods from Chronoanalysis

Chronopotentiometry

The application of Chronopotentiometry could be derived into two parts. As an analytical method, the range of analysis is normally in the range of 10^{-4} mol/L to 10^{-2} mol/L, and sometimes it will be as accurate as 10^{-5} mol/L. When the analysis is in the extremely lower range of concentration, lower current density could be used. Also, to get the accurate concentration determination, the transition time could be extended. In this area of analysis determination, chronopotentiometry is similar to polarography. Waves that are separable in polarography is also separable in chronopotentiometry.

Chronopotentiometry is an effective method to study the mechanism of electrode mechanism. Different electrode will have different relationship between E and t in the chronopotentiometry graph. In this situation, E is the electrode potential in voltage and t is the reaction time in seconds. By the method of studying the relationship between E and t in the chronopotentiometry graph, we can get the information of a lot of mechanisms of electrode reactions, such as the electrode reaction of hydrogen peroxide and oxalic acid. The experiment of chronopotentiometry could be done in a very short time period, so it is a good method to study the adsorption behavior at the surface of electrode. By studying the chronopotentiometry graph of electrode after adsorption of iron ion, it is proved that the adsorption of platinum on iron ions exists. By studying the chronopotentiometry graph of platinum electrode adsorbing iodine, it is proved that the adsorption of iodine happening in the form of iodine molecules, not iodine atom.

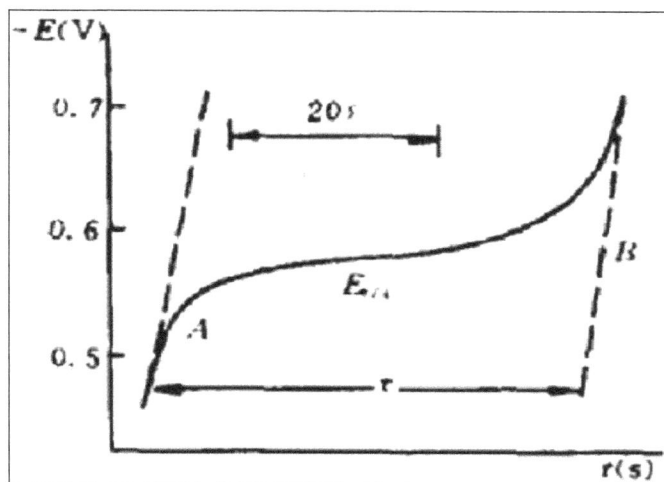

Chronopotentiometry.

Chronocoulometry

Chronocoulometry is an analytical method that has similar principle with chronoamperometry, but it monitors the relationship between charge and time instead of current and time. Chronocoulometry has the following differences with chronoamperometry: the signal increases over time instead of decreasing; the act of integration minimizes noise, resulting in a smooth hyperbolic response curve; and contributions from double-layer charging and absorbed species are easily observed.

POTENTIOMETRY

Potentiometry is one of the methods of electroanalytical chemistry. It is usually employed to find the concentration of a solute in solution.

In potentiometric measurements, the potential between two electrodes is measured using a high impedance voltmeter.

Use of a high impedance voltmeter in important, because it ensures that current flow is negligible. Since there is no net current, there are no net electrochemical reactions, hence the system is in equilibrium.

At its most fundamental level, a potentiometer consists of two electrodes inserted in two solutions connected by a salt bridge. The voltmeter is attached to the electrodes to measure the potential difference between them.

One of the electrodes is a reference electrode, whose electrode potential is known.

The other electrode is the test electrode.

The test electrode is usually either a metal immersed in a solution of its own ions, whose concentration you wish to discover, or a carbon rod electrode sitting a solution which contains the ions of interest in two different oxidation states.

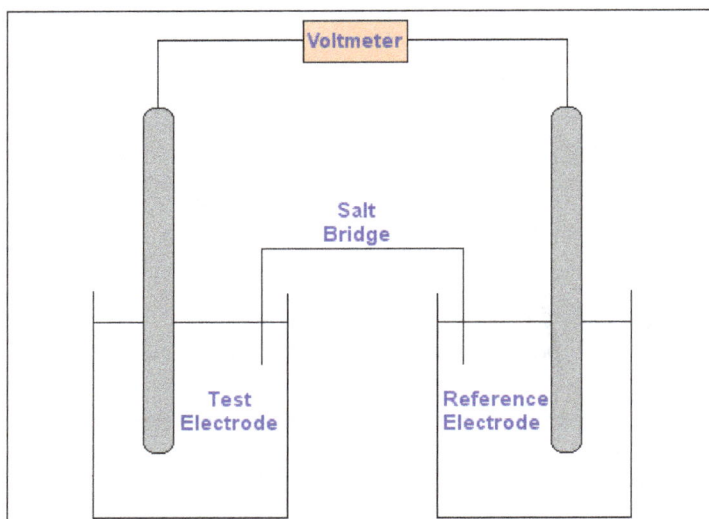

ELECTROGRAVIMETRY

Electrogravimetry is a method for the separation of the metal ions by using the electrodes. The deposition takes place on the one electrode. The weight of this electrode is determined before and after deposition. This gives the amount of the metal present in the given sample solution.

Principle

The main principle involved in this method is the deposition of the solid on an electrode from the analyte solution.

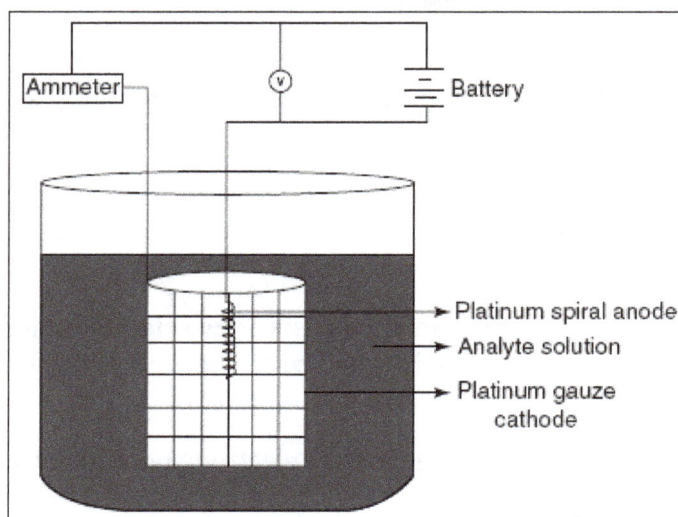

Electrogravimeter

The material is deposited by means of potential application. The electrons are transported to electrode by the following mechanisms:

- Diffusion

- Migration

- Convection

Theory

A metal is electrolytically deposited on the electrode by increasing the mass of the electrode.

$$M^{+2} + 2\,e^- \rightarrow M(S)$$

Therefore,

$$E_{electrolysis} = E_{cathode} - E_{anode}$$

The electrons deposition is governed by Ohm's and Faraday's laws of electrolysis which states that the amount of the electrons deposited on the electrode is directly proportional to the amount of the current passed through the solution and the amount of different substances deposited is directly proportional to the molar masses divided by the number of electrons involved in the electrolysis process.

That is the current (I) is directly proportional to the electromotive force *(E)* and is indirectly proportional to the resistance (R).

$$E = IR$$

From the above equation, we get the following:

$$E_{electrolysis} = E_{cell} - IR$$

$$E_{cell} = E_{cathode} - E_{anode}$$

Therefore,

$$E_{applied} = E_{cathode} - E_{anode} - IR$$

$$I = \left(-E_{applied}/R\right) + 1/R\left(E_{cathode} - E_{anode}\right)$$

$$I = \left(E_{cell} - E_{applied}/R\right)$$

$$I = \left(-E_{applied}/R\right) + K$$

where *K* is the constant.

A plot of the current of the applied potential in an electrolytic cell should be straight line with a slope equal to negative reciprocal of the resistance.

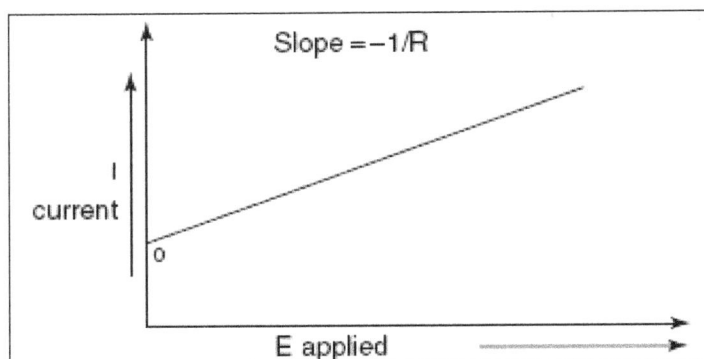
Electrogravimetric plot.

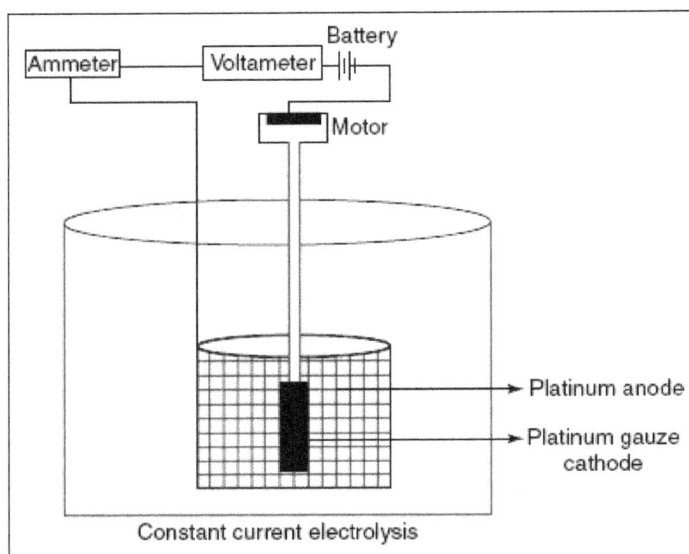
Constant current electrolysis

Types of Electrogravimetry Methods

There are mainly two types of electrogravimetry methods:

- Constant current electrolysis: By name itself it indicates that the constant current is applied for the electrons deposition. The instrument is composed of a cell and direct current source. A 6 V battery, an ammeter and a voltameter are used. The voltage applied is controlled by a resistor.

- Constant current electrogravimeter.

Factors affecting the Deposition

- Current density.

- Temperature.

- The presence of complexing agents.

- Chemical nature of ion.

Applications of Constant Electrolysis

Analyte	Conditions required
Ag^+	Alkaline cyanide solution
Cd^{+2}	Alkaline cyanide solution
Cu^{+2}	Acidic solution
Ni^{+2}	Ammonical solution
Mn^{+2}	HCOOH solution

Constant potential electrolysis: In this method, the potential of the cathode is controlled. It consists of two independent electrode circuits that are connected with a common electrode. For the better results, three electrode systems are used.

- Working electrode: Used for the deposition of the sample.

- Counter electrode: Used as a current sink.

- Reference electrode: Maintains the fixed potential despite the changes in solution components.

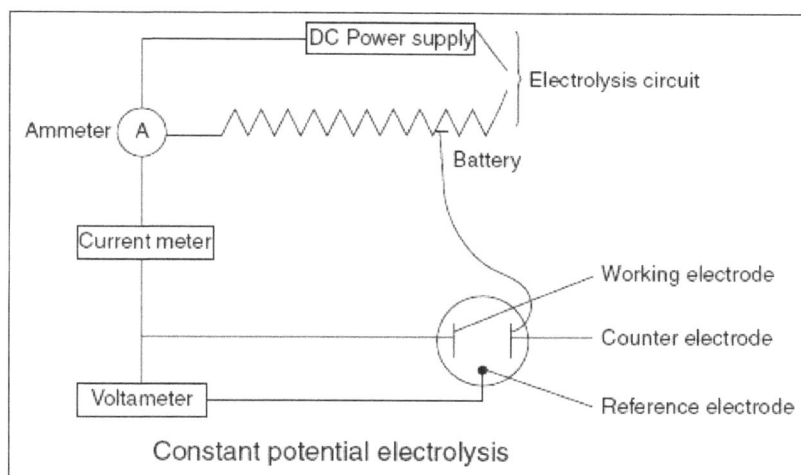

Constant potential electrogravimeter.

Electrode used in the Electrogravimetry

The electrode used in the electrogravimetry should posses the following characters:

- It should be non-reactive.

- It should be readily ignited to remove the organic matter.

The commonly used electrode is the mercury cathode electrode for the deposition. This has the following advantages:

- It forms the amalgam with number of metals.

- It has a high hydrogen voltage.

In this method, the precipitated elements are dissolved in the mercury. This method is mainly used in the removal of the reduced elements.

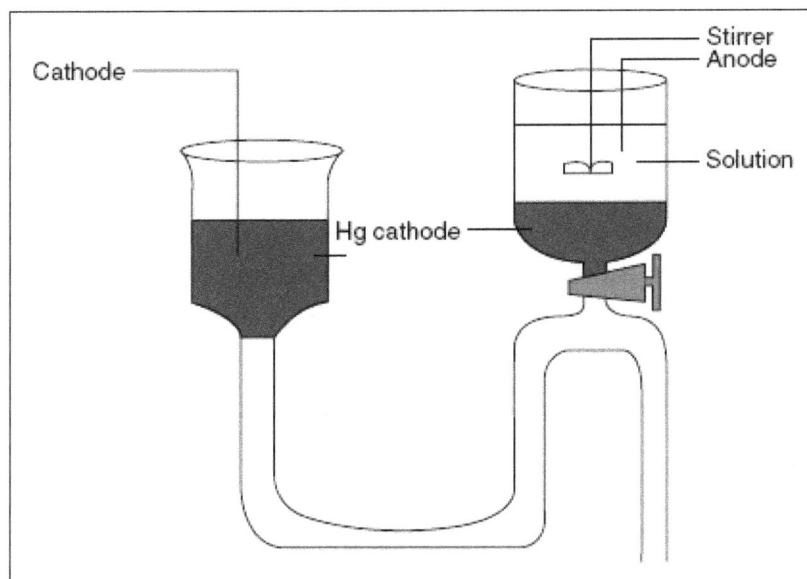

Electrodes.

Applications

- Used in the successive deposition of the metals.

- Example: Cu, Bi, Pb, Cd, Zn and Sn.

- Used in the simultaneous deposition of the metals.

- Used in the electro synthesis.

- Used in the purification process by removing the trace metals from the samples.

References

- Electroanalytical-method, materials-science, topics: sciencedirect.com, Retrieved 19 April, 2019

- Fritz Scholz (21 December 2013). Electroanalytical Methods: Guide to Experiments and Applications. Springer. Pp. 109–. ISBN 978-3-662-04757-6

- Coulometry: blogspot.com, Retrieved 12 July, 2019

- Laborda, Eduardo; González, Joaquín; Molina, Ángela (2014). "Recent advances on the theory of pulse techniques: A mini review". Electrochemistry Communications. 43: 25–30. Doi:10.1016/j.elecom.2014.03.004. ISSN 1388-2481

- Stripping, Stripping, Techniques, EC-epsilon, manuals: basinc.com, Retrieved 29 June, 2019

- Douglas A. Skoog; F. James Holler; Stanley R. Crouch (27 January 2017). Principles of Instrumental Analysis. Cengage Learning. Pp. 658–. ISBN 978-1-305-57721-3

- Nahir, Tal M.; Clark, Rose A.; Bowden, Edmond F. (2002). "Linear-Sweep Voltammetry of Irreversible Electron Transfer in Surface-Confined Species Using the Marcus Theory". Analytical Chemistry. 66 (15): 2595–2598. Doi:10.1021/ac00087a027. ISSN 0003-2700

- Polarography, science: britannica.com, Retrieved 19 April, 2019

- Francis George Thomas; Günter Henze (2001). Introduction to Voltammetric Analysis: Theory and Practice. Csiro Publishing. Pp. 58–. ISBN 978-0-643-06593-2

- Copeland, T. R.; Skogerboe, R. K. (2008). "Anodic stripping voltammetry". Analytical Chemistry. 46 (14): 1257A–1268a. Doi:10.1021/ac60350a021. ISSN 0003-2700

- Potentiometry, definition: chemicool.com, Retrieved 11 January, 2019

PERMISSIONS

INDEX

A

Alkaline Fuel Cell, 100, 102, 106

Aluminium Alloys, 59, 146-147

Anode, 3-9, 11-12, 24-25, 28, 34, 38, 41, 45-46, 56-57, 59-60, 66, 68, 70-71, 90, 92, 108, 124, 127, 136, 140, 143, 146, 148, 162, 168, 173, 177, 179, 186, 192, 198, 216, 225

Anodic Index, 70-71

Anodizing, 59, 83, 91, 146-153, 193

Ascorbic Acid, 38, 73

B

Benzotriazole, 73

Black Oxide, 83-85, 87-88

C

Calcium Carbonate, 18, 49

Calcium Oxide, 49, 65

Carboxylic Acids, 99, 181

Castner Process, 133, 153-154

Cathode, 3-9, 11-12, 24-25, 28, 34-35, 38, 41, 43, 45-46, 59-60, 66, 68, 70, 91-92, 99, 101-108, 122-124, 129, 141, 145, 148, 162, 168, 171, 173, 179, 187, 192, 210, , 227

Cathodic Protection, 37-38, 56-57, 59-60, 66, 69-70

Cell Notation, 11-12

Chloralkali Process, 133, 156-159, 193

Chlorine Gas, 2, 11, 136, 138, 155, 157-158, 167-168, 192

Chromate Conversion Coating, 91, 148-149

Chronoamperometry, 30-31, 202, 219-221, 223

Combination Reaction, 27, 48-50

Compressed Hydrogen, 102, 109, 168, 172, 176

Conversion Coating, 83-86, 91, 148-149

Copper Selenide, 88

Copper Selenium, 83-84

Corrosion Inhibitor, 72, 80

Corrosion Reaction, 27, 51, 60

Cottrell Equation, 23, 26, 212, 220

Coulombic Force, 143

Crevice Corrosion, 53, 55

Cyclic Voltammetry, 30, 201, 208-216, 221-222

E

Electrochemical Cells, 1, 3, 5-8, 24, 38, 95, 98, 167, 169, 220

Electrochemical Reaction Mechanism, 30, 202

Electrode Potential, 7, 23-25, 29, 34, 60, 92, 165, 167, 205, 210-211, 217-218, 221-223

Electrolytic Cells, 2, 6-8, 10, 25, 135-136, 164

Electrolytic Passivation, 146

Electroplating, 1, 38, 70, 83, 90, 94-95, 126, 135, 177, 182, 186, 190, 193, 218

Equilibrium Constant, 14-18

Esters, 16, 75

F

Ferrous Hydroxide, 72

Fume Bluing, 86-88, 90

G

Galvanic Cells, 2-3, 6-8, 12-13, 25

Galvanic Corrosion, 65-69, 71

Galvanic Couple, 55, 58, 66

Gibbs Free Energy, 25, 43, 171-172, 174, 181

H

Hexavalent Chromium, 91-92, 94

Hofmann Voltameter, 139-140, 168

Homogeneous Equilibrium, 14-15

Hydrobromic Acid, 122-123

Hydrogen Evolution Reaction, 99, 171, 173

Hydrogen Fluoride, 36, 161

Hydronium, 64, 72, 160, 166

Hydroxide Ion, 3, 192

I

Iron Oxides, 37, 52, 56, 80

L

Lithium-ion Battery, 115-116

M

Manganese Dioxide, 4, 27, 88

Mass Transfer, 18-23, 200, 208

Microbial Corrosion, 56, 73

Molar Flux, 20

N

Neodymium Magnets, 57, 61

Nernst Equation, 4, 12, 14, 43, 120, 166, 201

Nickel Catalyst, 34

Niobium, 55, 60, 146, 151
Nitric Acid, 38, 79, 87-88, 146

O
Oxidation-reduction Reaction, 13, 47, 135
Oxygen Evolution Reaction, 110, 173

P
Phosphoric Acid, 52, 101, 103, 106, 109, 150
Potassium Hydroxide, 4-5, 75, 79, 102-103, 107, 167
Potassium Nitrate, 7, 87-88
Proton Exchange Membrane, 101, 105, 115, 124, 170, 172

R
Redox Cycling, 39
Redox Reaction, 8, 25, 32, 34-36, 38, 45, 47, 71
Rust Bluing, 86-88

S
Sodium Nitrite, 66, 152
Sodium Silicate, 62, 65, 151-152

Solid Oxide Fuel Cells, 101, 107, 111
Standard Hydrogen Electrode, 7, 24, 34, 41-43
Stress Corrosion Cracking, 53
Sulphur Dioxide, 15, 48
Sulphuric Acid, 50, 209

T
Thermal Stress, 146, 148
Trivalent Chromium, 92

U
Ultramicroelectrodes, 31, 213

V
Voltaic Cells, 6-7, 9, 135-136

Z
Zinc Dithiophosphates, 73-74, 76

www.ingramcontent.com/pod-product-compliance
Lightning Source LLC
Chambersburg PA
CBHW061248190326
41458CB00011B/3612